T0311050

A Climate Policy Revolution

A CLIMATE POLICY REVOLUTION

What the Science of Complexity Reveals about Saving Our Planet

ROLAND KUPERS

Harvard University Press

Cambridge, Massachusetts
London, England
2020

First printing

Library of Congress Cataloging-in-Publication Data
Names: Kupers, Roland, 1959– author.
Title: A climate policy revolution : what the science of complexity reveals about
saving our planet / Roland Kupers.
Description: Cambridge, Massachusetts : Harvard University Press, 2020. |
Includes bibliographical references and index. | Summary: "In this book, Roland Kupers
argues that the climate crisis is well suited to the bottom-up, rapid, and revolutionary
change complexity science theorizes; he succinctly makes the case that complexity science
promises policy solutions to address climate change"—Provided by publisher.
Identifiers: LCCN 2019054773 | ISBN 9780674972124 (cloth)
Subjects: LCSH: Complexity (Philosophy) | Climatic changes—Government policy.
Classification: LCC Q175.32.C65 K86 2020 | DDC 363.738/74561—dc23
LC record available at https://lccn.loc.gov/2019054773

CONTENTS

A Climate Policy Revolution

Time's Up

Nothing is so painful to the human mind as a great and sudden change.

MARY WOLLSTONECRAFT SHELLEY, *FRANKENSTEIN*

We procrastinated for the best reasons. We weren't sure. We didn't believe—or hear—those who said they were sure. Some climate scientists overstated how sure they were. The media reported that economists thought it would be expensive to deal with. Their equilibrium models told us so. Other economists said there was an issue with how we value the future, but it would still cost a lot of money. Other people didn't realize that carbon dioxide actually accumulates in the atmosphere—and doesn't just disappear. Some heard Al Gore expound his inconvenient truth, and then they felt like someone should do something—or that he was pontificating too much. And other things seemed more urgent. Like poverty alleviation, loss of biodiversity, gender issues, or the next mortgage payment. In any case we procrastinated and were confused.

The 2018 report of the International Panel on Climate Change (IPCC) trumpeted that time is up. It is now or never—stated with *high confidence*.[1] Summarizing over 6,000 science papers that describe the state of the climate crisis, the IPCC laid it out one more time.[2] One United Nations official described it as "a deafening, piercing smoke alarm."[3] Never before has an institution been created to establish the scientific consensus at such a scale. The IPCC pores over all known science on our collective behalf, deals with contradictions, and then articulates our state of knowledge and of uncertainty. It is not perfect, but it is the most sustained effort ever undertaken. In its comprehensive, scientific, and transparent approach it is a huge improvement on Galileo's trial, the witches of Salem, or even the Mueller investigation. But the alarm may still fall on deaf ears.

The unambiguous conclusion is that there is a last chance—but only just—to keep global warming to 1.5 degrees Celsius.[4] The difference between

1.5 and 2 degrees? Coral reefs, for example. In one case we keep them, in the other we don't. Also, there is *high confidence* that nature will maintain more of its services to humans—to us. That means fewer new diseases and droughts, better crop yields, fewer flooded coastlines, greater biodiversity. Allowing the global temperature to rise by 2 degrees could double the losses in annual ocean fish catches, up the number of people exposed to water stress by 50 percent, and increase the declines in the yields of key staple crops such as maize, rice, and wheat.[5] In short, a 1.5 degree scenario is not just 0.5 degrees less than 2 degrees; the damage does not have a linear relation with the temperature; it increases vastly with the additional half a degree. Reacting to people's perception about their credibility, the IPCC over time has developed a real vocabulary of uncertainty. *High confidence* is their gold standard, equating to confidence between 90 and 100 percent. If there is a 90–100 percent chance of your car brakes failing, or a typhoon coming your way, you would act accordingly. So now there is no time left to procrastinate. Must fix the brakes, shelter from the typhoon, but how?

Psychologists tell us that this approach is exactly the wrong way to motivate people to take action.[6] Describing disasters is a good way of attracting people's attention, but it is a poor way to sustain personal engagement and actually get people to do something.[7] Scaring people doesn't work. However, the idea here is not to convince the reader at this point that we must do something about the climate crisis, but simply to remind ourselves of the facts. There are huge positive aspects to dealing with the climate crisis. Not least of which is the opportunity to reinvent. Our world is far from perfect, so given the opportunity to change it, we can make it better in terms of well-being, health, or equity. We'll return to the issue of motivation and social norms in Chapter 8, but first the core idea of this book—revolution policy.

The climate crisis requires rapid and sustained change—starting now. And that implies sweeping systemic change, not just tweaking an existing equilibrium. Economics is great for equilibrium systems, or small deviations from it. But it struggles with systemic change. In fact, the IPCC used to commission a separate economic report as part of its work—not this time. Part of the world's dithering about climate change is because we've traditionally left most of the framing of a solution to economists. And the economics that underpins policy is hopelessly mired at the margin. Incremental change—in the long run—is its method. And with climate change, time has run out for gradual change. Forget the long run; in the short run some of us may in fact be dead. We need a revolution, not mere incremental change. So, economics needs to get out of the way of the discipline that can tackle this problem far more expeditiously, namely, complexity.[8] In principle, economics might have integrated complexity—but that also has proven to take too long.[9]

Changing Fast or Slow

When large-scale and rapid change is required, you are justified to think that the only way to achieve this is through robust top-down action, with a strong leader with lots of centralized power making big changes. It worked during World War II to jolt the economy onto a war footing. General Motors stopped producing cars and churned out tanks—turning on a dime, following the instructions of a president freshly empowered by wartime solidarity. It is not wrong that centralized power works, but in the case of climate change, it is simply unlikely to happen. Today's strongmen are just not interested. They care about power, about racial identity and maintaining privilege, less about the state of the planet. Such disconnect between authoritarianism and care for the earth is not a given. For all his many other faults, Fidel Castro ensured that Cuba has a remarkable environmental legacy.[10] President Xi Jinping has started to turn China's juggernaut on a greener course, even calling for an ecological civilization. But most of the current crop of strongmen don't care—or don't care enough.

Top-down action might have worked, but it won't happen—or won't happen soon enough. Most democratic societies have amply demonstrated that they are not capable of taking top-down actions that are proportionate to the severity of the problem. To some extent their governance systems were designed to avoid such drastic action, with checks and balances allowing various stakeholders to express their interests. Occasionally this leads to special interests capturing power, such as when the European car industry colluded to reverse the improvement in urban air quality through misleading claims about its diesel engines. But while some are tempted to blame special interests, the inability to take drastic action is more a feature of the system than the result of the collusion of dark forces.

And who should take the top-down actions in the first place? Generally the government. We tend to refer to government as if it were outside of the system. But it's not. It is inside, connected and locked in. So most of the time it can't take the actions that it would need to.

Bottom-up change has an undeserved bad name. It is often associated with ineffective plodding, people arguing all the way to the goal line, tripping over each other. Obviously that happens too, but a deeper reason for mistrusting bottom-up processes is that we have not been familiarized with how they actually work. How do systems self-organize? What is their source of order? This is where complexity science comes in. And when done right, bottom-up change can be quick and sweeping, with its impact proportional to the scale of the problem.

We don't need to give up on top-down action, just not make it the only game in town.

In 2018 Australia joined a growing number of countries—among them Ireland, France, China, and the Netherlands—that have adopted measures to discourage plastic bag usage through bans or taxes. Kenya took the most extreme steps, with penalties of up to four years in prison or fines up to $39,000. Uniquely, in Australia this led to selective customer rage. Reportedly, supermarket staff were abused, leading several chains to back down.[11] Contrast that with Ireland. In 2003 a small charge on bags was introduced and, unexpectedly, within weeks this led to a 94 percent drop in use.[12] A revolution had occurred. Both in Australia and in Ireland these outcomes were not expected.[13] Yet getting this kind of sweeping change is every policy maker's dream. But how can they do it more purposefully and effectively?

Climate science all but tells us we need a *revolution*, a state shift moving the world onto an entirely different plateau. That holds true almost regardless of the overarching climate target, or the "carbon budget" under such a target. It is certainly true for the Paris Agreement mentioning "1.5°C" as a threshold for average global warming above preindustrial levels. Such targets are difficult to achieve given the massive barriers to dramatically reducing carbon emissions. In the face of all of this, some resort to calls to abandon economic growth altogether and focus on shrinking the economy.[14] That is unlikely to work, as growth is needed to provide the oxygen for change. Those affected, such as the Appalachian coal miners, must be helped. Because it is the right thing to do, and also because they may otherwise help anoint leaders who slow things down.

How to reconcile the need for radical change, for a "revolution" with ambitious overall targets, with the familiar policy tools that appear unable to deliver what is required? Policy makers deal in smooth linear change, not in step change through purposeful revolutions. One important reason for this is that much of policy is grounded in the science of economics, at least in what economists refer to as "orthodox" economics. Such economic policy leads us to think at the margin—in terms of smooth incremental change. Economic models do not accommodate revolutions or discontinuities very well. Sudden change, such as the 2008 financial crisis, does not fit or follow from those models—even in hindsight.[15]

Let's note that any sweeping generalization of a discipline is unfair, and this applies to economics as well. Like any social science, the discipline itself is a complex system, with a great deal of diversity of perspectives and interacting views. Economists often place themselves on a range, from the orthodoxy of neoclassical views to the unorthodoxy of complexity. This is a fascinating story, but one that has been told elsewhere.[16] For the purposes of climate policy the issue is not with economics, where a broad range of views can be found. Both the economic models and the narratives that inform climate policy are overwhelmingly based on orthodox perspectives. Indeed, there is no shortage of leading economists who express dissenting views, such as Mariana Mazzucato, Martin Weitzman, Nicholas Stern, Doyne Farmer, Geoffrey Heal, and others.[17] The problem is that these views have little influence on climate action, which is mostly grounded in a more standard, more orthodox economics. The diversity of views among economists does not subtract from the fact that orthodox economics has had a profound influence on the framing of climate policy—a framing that has led us astray.

Imagine that you desire a sweeping change. Are you confident that an incremental pathway gets you all the way to your goal? Or is there a real possibility that your ultimate goal remains forever out of reach, because the pathway you chose is simply not steep enough? If you conclude that you need a revolution, are you condemned to resort to the traditional kind, bracing for war and plotting to overthrow the establishment? That is not the case—at least not always. When do incremental change and revolution coincide? When and how do they differ? What do we know about their similarities and differences? How can you be purposeful about getting the kind of change that shifts an entire system to a different state?

While quick change is enticing in general, for the climate crisis it is imperative. And history shows that rapid change in societal systems is possible. Therefore, this book does not call for more stabs at gradual change. Rather, it calls for peaceful, system-wide resolution through policy. We've done this before, and must do it again. Some refer to it as the third, fourth, fifth, or even sixth revolution.[18] However you view it, the key here is that revolution is not only called for, it is required, and we can do this.[19]

Incremental Change or Revolution?

Stare into a pot of water being heated, and you see very little—until suddenly, and seemingly instantaneously, it boils. Physicists long ago described the laws that underpin this. They call it a "phase transition." It is intensely familiar yet should also be a source of wonderment. Why is it sudden? Which "switch" flips to make it happen?

Similarly sudden transitions occur in nature, for example, the proverbial butterfly effect, where a seemingly small change can lead to a system-wide effect: a flap of its wings in the Amazon causing a hurricane a continent away. The reintroduction of wolves in Yellowstone National Park, so the story goes, ultimately led to changes in the flows of rivers, via a long cascade of connections. The wolves changed the grazing habits of the elks, who had become accustomed to meals of young cotton trees and willows, uninterrupted by predators. But beavers need willows to survive in winter, so their numbers dropped. Now that the elk diversified their grazing behavior, the beavers flourished and built dams that changed the course of rivers. But as in any complex system, even this is a simplification, and the full dynamics are yet to be told. The rivers changed through a systemic causation, not from the direct cause of reintroducing the wolves.[20]

Social systems, too, can stumble across a threshold: simmering local discontent sometimes boils over suddenly and leads to a revolution.[21] Often we overemphasize a single precipitating event—one small thing that leads to a cascade of others. The butterfly of World War I was the assassination of Austria's Archduke Ferdinand by a nineteen-year-old Gavrilo Princip in 1914. Four years later, sixteen million people had died. The war itself involved over one hundred nations and laid the groundwork for the discontent leading to the rise of Hitler and World War II, which, in turn, led to the Cold War, and so history rolls on.[22] In reality, dramatic shifts such as wars have many causes. Many interwoven factors poise the system for change, but it is the trigger vent that people recall as the start—or the catalyst.

Was any of that predicted? Certainly not universally. A year before the assassination, *The Economist* famously penned an editorial titled "Neighbours and Friends." It argued how "slowly but surely . . . war between the civilized communities of the world" was becoming "an impossibility."[23]

Political revolutions happen suddenly, and there are many other discontinuities in our world: new products sweep the world, as do ideas large and small. Some changes happen suddenly, but many others take forever, as vested interests are often too strong, effectively gluing the system in its place. Niccolò Machiavelli captured the sentiment perfectly in *The Prince*: "There is nothing more difficult to take in hand, more perilous to conduct, or more uncertain in its success, than to take the lead in the introduction of a new order of things. Because the innovator has for enemies all those who have done well under the old conditions, and lukewarm defenders in those who may do well under the new."[24]

Machiavelli understood that even if the ruler had the power to change things top-down, the backlash would unravel or dull much of the initiative.

Social change can be slow, at least to those wanting to see it happen. Civil servants or leaders of nongovernmental organizations who are in the business of changing society can only dream of a sudden sweeping change. Much of the time, things stubbornly refuse to budge. Reducing traffic accidents, corruption, or greenhouse-gas emissions is fiendishly hard. There is typically not a single silver bullet. It is often more akin to a silver buckshot and seeing what sticks.

The science of system-wide change is, in fact, often more art than science. That leads to two fundamental reactions. Some say that large systemic changes are inherently hard to predict, so why even try? Others say that not keeping systemic change in perspective misses critical opportunities.

Complexity—The Science of Interconnected Systems

Enter what, over the past thirty-odd years, has come to be known as "complexity" or "systems science."[25] Catalyzed by the Santa Fe Institute, the science of complex systems is starting to shed new light on the different evolutions of a system as a whole, whether that implies incremental or revolutionary change. The very word "complex" stems from the Latin *plexus*, for braided or interconnected; so complexity is derived from something that is "with braids." It is, quite simply, the science of interconnected systems. Now other aspects become important. Complex systems, by definition, are highly interconnected and constantly changing; networks and interactions are central to the thinking.

How do complexity and economics relate? For one, complexity provides some models that start getting to grips with sudden system-wide change, and economics doesn't. But pitting the two disciplines against each other isn't exactly fair. Most classical economists—those in the eighteenth and nineteenth centuries—were very comfortable with complexity.[26] Adam Smith strived to distill human desires and the workings of economics and society down to the most important drivers, while also putting complexity issues such as moral sentiments front and center. Many other economics thinkers, from Joseph Schumpeter to Thomas Schelling to contemporaries like Daron Acemoglu, have followed in this tradition.[27]

The differences between economics and complexity science can be overstated. It is true that many economists see complexity scientists as having good critiques and little else.[28] It is also true that some complexity scientists have the view that much of economics is lost in streamlined math based on erroneous—or at least severely limiting—assumptions. But beyond the polemics, and because the two approaches share much more than mutual disdain, an understanding of complexity is imperative to trigger policy revolutions—in particular to deal with climate crisis.[29]

A Complexity Lens

The traditional economic framing of a policy area such as the climate crisis holds the danger that an approach is chosen that does not fit the underlying complexity of the system. As a consequence the policy may be inefficient or simply ineffective. This is a lesson that also holds for other policy areas. There are many examples of failure in development projects that assume an oversimplified system and take a top-down intervention approach.[30] Shipping food aid into a famine area threatens to destroy the remaining food businesses by outcompeting them. When considering the energy transition to a low-carbon system, the energy system itself is treated as a relatively simple system, one that can be framed in terms of fairly linear dynamics. Taking a reductionist perspective, the behavior of the system is the sum of the behavior of its various components. Direct causality applies. That in turn means that the outcome of a particular intervention can be predicted and understood.

A complexity lens on the energy system would consider its deep and manifold interconnections with other societal systems: promised pension payments are critically dependent on the superior returns of fossil fuel companies; the attachment to coal production in Appalachia, the Ruhr, or Poland is as much cultural as it is economic; risk models in the financial sector tend to overestimate the vagaries of new technologies; people resist government's intervention to make their homes smaller energy guzzlers. We could go on. The energy system maybe at the heart of dealing with the climate crisis, but it is not isolated. Like a ball of badly cooked spaghetti, it sticks to its surroundings: it is part of a complex system.[31]

This Book

In Part I we'll explore some complexity ideas. Just like people, systems carry their past with them and are bound by historic ties. Such path dependencies must be recognized and sometimes purposefully tweaked. We look at the growing knowledge of networks and how different types of networks lead to profoundly different outcomes. We've briefly alluded to the assumption of equilibrium in economics; exploring the consequences of systems being either in or far from equilibrium opens up a rich vein of thought and insight. And finally we return to the ideas of top-down versus bottom-up change. Bottom-up dynamics are at the heart of what makes successful societal systems work.

In Part II we turn to a few examples of revolution policies. We don't have the magical formula for fixing the climate instantly, but these examples illustrate how different the perspective is from the bottom up, and how it just might unleash a revolution. Coal is the hardest nut to crack and appears to yield only to flanking maneuvers. A climate-friendly future cannot happen with the social norms we have today. If billions more join the middle class as it is defined today, we are quite literally toast. The very idea of the middle class must morph. The good news is that social norms don't exist in isolation and that they change through context. We'll explore how. In matters of climate, business is in turn maligned or hailed as the savior. We cut through the confusion and describe the contribution and the limitations of business to come up with new approaches. One particular aspect of innovation concerns widespread misunderstanding about its origin and how it is paid for. We look at autonomous vehicles, not as transport solutions, but as climate catalysts.

Finally we attempt to provide some recipes for the kind of revolution policy that is required to face down the climate crisis.

PART I

COMPLEXITY IDEAS

(2)

Getting Unstuck

Part of company culture is path-dependent—it's the lessons you learn along the way.

JEFF BEZOS

In July 1907 a sleek and quiet bus from the London Electrobus Company picked up its first passengers in London. The bus was one in a fleet of twenty battery-powered vehicles. The company had entered into head-on competition with the loud, smelly, and unreliable buses powered by internal combustion engines. Those had been introduced in the previous year. In this early period, the future of public transport technology was up for grabs. The electric buses were well designed, with a range of sixty kilometers. They stopped along their route at a garage in Victoria, where the depleted batteries were swapped out with fully charged ones in a mere three minutes. Because of the regular routes of a public transport bus, this limited range presented few operational inconveniences. Electrobus held all the cards. The future of public transportation appeared to be electric, and at the time London was an important beacon for the world.

Alas, it was to remain the largest electric bus experiment of the entire twentieth century, as it also became one of the first major corporate scandals of the same century. The shareholders, German lawyer Dr. Edward Ernest Lehwess and the colorful Baron de Martigny, through a network of companies, siphoned off funds in order to live the high life on the Côte d'Azur.[1]

The company collapsed in ignominy. *The Economist* concludes: "Whether the fraud was truly a tipping point for electric vehicles is, of course, impossible to say. But it is a commonplace of innovation—from railway gauges to semiconductors to software—that the 'best' technology is not always the most successful. Once an industry standard has been established, it is hard to displace. If Lehwess and Martigny had not pulled their scam when they did, modern cities might be an awful lot cleaner."[2]

An obvious, arguably superior, and expected path suddenly changed through a relatively small and unexpected development. One potential revolution was thwarted, and another was enabled: small interventions can have enormous consequences.

In this book we are after the opposite of what the London Electrobus Company accomplished: purposeful revolutions. How to identify the small interventions in a system that might trigger rapid and sweeping change? This is useful when we are in a hurry to get something done, such as getting rid of smoking—or necessary when we have run out of time, such as with climate change. Sometimes it is just desirable to go faster, and other times it is just necessary. It is simply better to get rid of plastic bags quickly, but it is imperative to reduce carbon emissions very quickly.

Small Cause—Large Effect

Systems have a habit of getting stuck on a particular path that is difficult to dislodge. That is really the first thing to remind ourselves of when designing climate policy. It will often take much more than a mere price signal to shift incentives in the economy. Getting to the bottom of the reasons for a system having gotten stuck in a particular state is far from obvious.

When you have to lift a heavy weight, you call for a big crane. When the price of crude goes down, we assume there is an oversupply of oil. A common assumption underlying such statements is that there is proportionality between cause and effect. To get a big change, you need a big effort: climate policy requires huge changes in society, so it is a huge effort. The degree of change involved in dealing with global warming is daunting, implying an immense amount of work, investment, and mindshare. Perhaps not, or not always.

Complex systems have nonlinear causality. Small local tweaks can lead to system-wide effects. A small event (and a small mind) led to the elimination of electric buses at London in the beginning of the twentieth century. This is not a mere anecdote; it is a fundamental property of complex systems, that small causes can have large effects.

Segregation

In the 1960s Thomas Schelling delivered a remarkably simple explanation for the fact that neighborhoods in the United States tended to be highly segregated along racial lines. Small personal preferences for the ethnicity of your neighbors can lead to broad, macro-scale effects such as largely seg-regated neighborhoods.[3] Schelling demonstrated this masterfully in one of the earliest models capturing the dynamics of sudden change.[4] He made the first snapshot of the internal workings of a revolution—the first model of a complex system. Reportedly using pennies and dimes on a board rep-resenting two ethnic groups, he played out the influence of those prefer-ences (he was careful to describe the coins in his papers as 'colored chips' to avoid any suggestion of a value difference). Surprisingly, what happens is that a mild preference to live among your own ethnicity becomes ampli-fied, as it were, and leads to a near-total segregation of the neighborhood. Even desiring only a third of one's neighbors to be of one's own ethnicity triggered a revolution: it changed the neighborhood from a mixed one to a segregated one. Once the separation of communities is in place, the pat-tern becomes frozen in place and impossible to displace. Impossible, that is, without a major external intervention.

The system, in short, becomes enslaved to its history and the path it has taken. Path dependence, it turns out, is an important feature of many sys-tems.[5] Path dependence is like an addiction, a behavior that is difficult to break, except that it is not an individual who is addicted, but a system. It is systemic addiction. Schelling's model illustrates how this dynamic plays out for segregation. Any model that does not include such path dependen-cies and simply assumes that the momentary forces at work can shift the system is doomed to show gradual change only, or at least doomed to guide policy in the wrong direction.

Schelling's modeling confirmed what urban developers had known intui-tively all along: once segregation is in place, it is extremely difficult to dis-lodge. The task for urban planners, then, is to avoid this kind of undesir-able lock-in in the first place. Singapore successfully enforces a defined ethnic mix in each of its public housing buildings. This is undeniably effec-tive but is not an option everywhere. Short of forcing people to live in cer-tain neighborhoods, what's there to do?

Oak Park, a Chicago suburb, has been held up as a prime example of a place that got racial integration right.[6] Oak Park didn't start out that way. Like many neighborhoods in and around Chicago and other major cities, it too was racially homogenous. In Oak Park's case, it had been white and became mixed, and the intent was to keep it that way.

It turned out that a strong driver leading to segregation was the fear on the part of white homeowners that their property would lose value with blacks moving in. This worked to amplify the mild ethnic preferences that people might hold and get people to actually move house. The set of direct interventions that often gets the most credit for diversifying Oak Park consisted of two seemingly innocuous policies: an "equity assurance program" that keeps homeowners whole in case their property loses money, and a ban of for-sale signs. The latter was meant to ensure that any individual, possibly unrelated sales wouldn't start a larger trend and trigger the tipping point identified by Schelling.

Does this really totally explain the avoidance of segregation? There have been precisely zero claims under Oak Park's equity assurance program, and the ban on for-sale signs is undoubtedly unconstitutional but has never been challenged in court by realtors.[7] One explanation might be that the simple availability of these options has done its job. Knowing that the assurance program exists is assurance enough. No one even needed to use it for it to work.

There have been additional measures that have influenced the process. The equity assurance program and sign ban are only two small interventions, part of a much larger set. Oak Park has a Village Board of Trustees, which, in turn, gets input from twenty-five citizens' advisory commissions. There's a Housing Policy Advisory Committee, the Oak Park Regional Housing Center, the Oak Park Residence Corporation, the Oak Park Housing Authority. Some responsibilities are clear. Others aren't. What is clear is that the full intervention consists of a complex set of players no simple economic model can capture. There were multiple interventions by a great many actors, but those actions were powered by a loosely shared vision and direction.

None of this nullifies the attempts by Schelling and others to come up with models capturing the most salient of facts. The models cannot capture everything that is at work, but they do show the essence of what triggers this complex system to move from one state to another. What is clear is that the models for revolutions need to capture apparent "quirks" in human behavior, showing how little things can add up to a whole lot.[8] One small step in that direction is to avoid assuming that everyone is the same.

Bury the Representative Agent

Models can help us make sense of complex reality. But in simplifying, we lose something. What good is a map if it is as big as the landscape it purports to show? The big test of any model is whether it can serve as a useful guide to actual human and societal behavior—not necessarily to predict, but to provide insight.[9]

The idea of the representative agent—a model of the behavior of a single human being—is at the core of economic thinking and is a consequence of much standard policy thinking.[10] Without it, little economic modeling would proceed.[11] Step one: Translate fundamental human desires and decisions into axioms of one representative human being's behavior. Step two: Use that one agent's behavior to model how they all collectively behave in society. How else to make the decisions of millions or billions of people tractable?

Humans have, on average, one ovary (or one testicle).[12] That is a fact, but designing policy based on such a representative agent for humans is clearly absurd. Yet it shows how misleading it can be to average differences. Averaging across the properties of agents in a system erases the very diversity that is essential for the system to thrive. Allowing for heterogeneity among agents is an essential driver of bottom-up organization and learning in complex systems. In fact, diversity across agents is a core element in many other parts of economics: trade only happens because people want different things.

Models or policies with one representative agent, then, often miss important interactions. The rich and the poor behave differently, especially when facing each other. Humans don't just make purchasing decisions in a vacuum. They look to others for guides. "Keeping up with the Joneses" is real and ever-present. In principle economists know how to model such behavior—add the neighbors' wealth and consumption choices to one's utility function—but doing so complicates the math significantly. Hence, it is most often dropped in a standard framework with one representative agent.

Microeconomists generally have made good progress modeling such apparent quirks in human behavior, showing how little things can add up to a whole lot.[13] But in a review on the concept of representative agents, Alan Kirman concludes that "it is clear that the 'representative' agent deserves a decent burial, as an approach to economic analysis that is not only primitive, but fundamentally erroneous."[14] In order to make systemic change happen, this burial becomes essential.

Complexity models take into account the behavior of every single individual and their differences. They sidestep the mathematical difficulties that have stumped economists, because they use a different tool. Like Schelling with pennies and dimes, and today on a computer, they simulate a system rather than trying to capture it in equations. Embracing the difference in how people react in different contexts is the key to capturing the mechanisms that make revolutions possible. Economics by and large does not do that; complexity does. To enable nonlinear change, policy makers need to expand their foundation to include complexity.

Learning from Bombers

The production of bombers in World War II was the first time that cost reduction from learning by doing was properly documented. Today the same dynamic plays out again with solar photovoltaic (PV).

The Willow Run manufacturing complex in Michigan was built for the production of B-24 bombers. It was large. The assembly line was a mile long, exceeding by half Tesla's Gigafactory. Starting well behind, the United States managed to produce almost three times as many bombers as Germany over the period of the war. At its peak, Willow Run produced over 600 aircraft a month to support the Allied war effort.[15] But it mightily struggled to reach peak capacity and did not do so before 1945, when its economies of scale were fully realized—so much so that it was sometimes referred to as Will It Run?

And building aircraft was a lot more complex than making either batteries or cars. Indeed, Ford Motor Company, which managed the operation at Willow Run, struggled to increase volumes, as its hallowed manufacturing methods proved too rigid. Henry Ford had originally offered President Franklin D. Roosevelt to build 1,000 planes a day, provided the design was frozen. Call it the "brute force" method of economics of scale. But it all depended on keeping the design constant. Meanwhile, the military understandably insisted that constant improvements were essential, as new lessons emerged every day. Changing design based on operational experience as well as due to material and labor shortages was essential to building good planes. Roosevelt rejected Ford's offer.

That things are cheaper to make when you make a lot of them is an old idea. In fact, Adam Smith had noted this centuries earlier when observing the manufacturing of pins. The rigid method of the car manufacturers could indeed get lower cost from scale of production, but did not have much of a learning system to deal with design changes.

Meanwhile, in Seattle, Boeing was building the vastly more complex B-17 Flying Fortress bombers with a very different approach, in many ways anticipating the flexible methods that were to catapult Japanese industry to global prominence in the 1970s and 1980s. B-17 manufacturing became the object of a classic economic study into learning by doing. Boeing climbed up the learning curve and slid down the cost curve. The first B-17E cost $242,200. The last B-17G cost less than $145,000, despite vast improvements from the first to the final model.[16]

Economies of scale are proportional to how much one manufactures at any given time. Learning by doing is cumulative to everything you have made in the past—typically a much bigger number, but only over time. Learning is based on experience and expertise, and its growth is faster than just from economies of scale. But learning by doing, too, comes with its own problems; it can keep even better ideas at bay by having too much of a head start.

Lock-Ins

Technological lock-in can be a substantial problem. Established technologies may become so comparatively ubiquitous or cheap that they can lock out potentially better technologies that no longer stand a chance to catch up.

The energy sector is rife with such examples. Solar photovoltaic technologies have been an unabashed success story. Their costs have plummeted. On average, prices have decreased by almost 25 percent for every doubling of total production.[17] But there are many different kinds of PV, made from different materials and technologies. The spectacular success of the current silicon PV technology might well be freezing out other potentially more efficient technologies such as thin-film solar or other promising ideas under development.[18] As new as the current solar panels appear to be, we may well already be locked into a design in such a way as to exclude further improvements. While the efficiency of converting photons to electrons rises roughly linearly in the lab, this only translates into newspaper headlines, not into better panels.[19] The higher-efficiency technologies are too far behind in the learning-by-doing race.

The success of silicon PV famously led to the bankruptcy of thin-film company Solyndra,[20] which was developing a competing design. Effectively, this was a race between new companies like Solyndra—who are innovating and learning by doing—with incumbents like traditional silicon solar panel manufacturers who are exploiting economies of scale. The innovators can win because they have the stronger cost reduction mechanism. The problem occurs when the incumbent has too much of a head start. This is plausibly the case for silicon PV. The technical term for this is path dependence.[21] Having established a successful and cost-effective product is an accomplishment. It could also imply being locked in within an inferior equilibrium.

The question for policy makers is whether it is their business to pay attention to such lock-ins and path dependencies—and actively consider breaking them at times. Innovation and new technologies are subsidized and encouraged all the time.[22] But are technologies deserving of support that would otherwise be excluded from the market simply because of lock-in also being subsidized?

Lock-ins come in different shapes and types. They can occur over time, as people build infrastructure and develop social norms or just habits that are hard to change. They can be based on economies of scale that require volume simply to become effective—like silicon PV. Natural monopolies are a direct consequence where one market participant—or one design—takes over the entire market. Learning by doing can create an even deeper barrier to change, as the cumulative experience builds and provides insurmountable barriers to entry, safe from external intervention.

Once the lock-in exists, changing a system incrementally will be hard. It takes a jolt, typically an external one, to move on to a new equilibrium and thus trigger revolutionary change.

Breaking Path Dependencies

While the future is open, it may be tempting to assume that the past is settled. Yet there can be plenty of different views on how path dependencies have shifted.

Take the case of U.S. emissions: energy-related CO_2 emissions have famously and unexpectedly declined by around 10 percent in less than a decade from the middle of the first decade of the twenty-first century onward.[23] This is a lot and exceeded reduction in Europe, where it was actually a prominent policy goal. Popular lore has it that this was largely due to the rapid switch from coal to natural gas. That switch clearly played a large role, though later work has ascribed as much to the rapid rise of renewables, with demand reductions accounting for most of the balance. Getting the account right is important, but all of that still leaves the question open of what has truly caused the decrease. Was it a confluence of lucky breaks? Was it deliberate?

The answer, as so often, is many-faceted. The availability of cheap shale gas clearly played a substantial role. A lot of credit goes to private investors like George Mitchell, who labored in plain sight but relative obscurity, hoping for the big breakthrough. But there too, deliberate government policies played a significant role, largely in the form of research subsidies.[24] Were those subsidies deliberate attempts to jump-start a revolution and break past path dependencies? Yes and no. Much of the research was foundational scientific work, far removed from the eventual revolution it sparked. Virtually nobody predicted that the big shift from coal to gas would suddenly occur around 2005 and in the few years that followed. In fact, before 2005 the oil and gas industry had made enormous investments to supply the United States with gas, in the firm conviction that it would become a large importer.[25] But there was another story.

Might the rapid decline in U.S. CO_2 emissions, to their lowest levels since the early 1990s, have had something to do with direct environmental interventions? Note that I say "environmental," not "climate." While comprehensive climate legislation fizzled and eventually died a slow, prominent death in the U.S. Senate, the same Senate instructed the U.S. Environmental Protection Agency to tackle mercury and toxic emissions directly.[26] The result? Dozens of coal plants needed to install expensive equipment to scrub their emissions and suddenly became uneconomic to run, aiding the switch to natural gas.

Add that to the early R&D subsidies for shale gas and to subsidies for renewables deployment, and it's hard to argue how active government interventions didn't play a role in the described rapid drop of CO_2 emissions in the United States. The real task, of course, is to try to foresee and plan these kinds of rapid changes.

Being purposeful about addressing path dependencies is important, but far from simple. While change may be messy, driven by a multitude of factors, there is still an underlying dynamic of evolution or revolution that can be discerned. The current path toward a rapidly warming climate is teeming with path dependencies and lock-ins. We've listed a few above, but the collection is vast. Australian habits of consuming large amounts of meat use vastly more energy than a plant-based diet would; the design of the U.S. sanitary systems that use enormous amounts of water and energy was locked in because of an antiquated conception of bacteria; and the recent oil money–fueled skyscrapers of Dubai have disregarded the traditional lessons on dealing with the heat of the desert. All these systems are locked into pathways that hinder dealing with climate change.

Getting On—and Off—Preset Paths?

The consolation when faced with lock-in systems is the knowledge that there may well be a small intervention that triggers system-wide change. This revolution approach is fraught with uncertainty, and the evolution path is more familiar. There are few sure bets. The alternative, however, is to pretend that these kinds of omnipresent path dependencies do not exist or constitute only minor factors of influence. What is the equivalent for major policy domains of suppressing visible for-sale signs to combat segregation? Which interventions are real? Which aren't? Which are necessary? Which are sufficient?

A further complication is that path dependencies don't play out in straightforward ways, but are part of various kinds of networks. It is through networks that we can get a better sense of how sudden nonlinear change can happen—the kind that is required to accelerate progress on climate action.

So we also need to look into networks.

③

Network Literacy

If a problem cannot be solved, enlarge it.

Dwight D. Eisenhower

The big idea from complexity science that we highlight in this chapter is our increasing understanding of the importance of network structures— and how it can help to shine light on path dependencies. Networks help us understand how small causes can have large effects and trigger the kinds of revolutions we need from climate policy.[1] We have already mentioned in Chapter 1 that complexity is about braided systems, or more precisely about complex adaptive systems.[2] Networks are a really powerful way of representing such systems. And insights into the dynamics of networks allow us to characterize systemic change. Instead of average and representative agents with averaged properties (think single ovaries), as standard policy often assumes, the very differences and connections between the agents become central to the emergent behavior of the system.

If your kids are not (yet) vaccinated for a particular virus, but they go to a school where everyone is vaccinated, they may still have some protection from the disease in question. This is because your kids are less likely to encounter another sick child, as all the kids around them are immune through inoculation. This is called herd immunity. Now, herd immunity for things like measles depends not only on average inoculation levels, but also on how the affected people are connected. Some people, known as anti-vaxxers, object to vaccinating their kids. We won't debate the dubious merit of their arguments here, but they are riding a wave of antiscience sentiment. It is plausible that most people who object to vaccination live and work in clusters, rather than being spread randomly through society. This is simply driven by people's slight preference to be around people like themselves. As we saw from Thomas Schelling's work, only a slight preference for the same type of people around you can result in a high degree of clustering.

So anti-vaxxers are likely to be clustered. However, if that is the case, epidemics are much more likely to develop in those unvaccinated clusters, even if the average number of anti-vaxxers across the whole region is low.

So you might get a breakout of disease inside some of these clusters of anti-vaxxers. Isolated cases happen all the time. This child is likely to infect a few other kids, although still contained within the cluster. Now, what happens if the network is such that the clusters are connected among each other, for example, if anti-vaxxers share a predilection for a particular sport that their kids play, across a bigger region, a sport that is part of the identity of their group? In that case the network structure will facilitate the spread of a broader epidemic, which will suddenly become visible, perhaps through frantic media reports. The whole phenomenon, however, depends on the structure of the underlying network. Vaccinating kids would obviously fix the problem, but if parents object, that may not be an option. In this example, it is the link between clusters through the sports activity that really triggers the broader pandemic. Perhaps discouraging the sports activity they share becomes a plausible alternative intervention? At the very least, anti-vaxxers should make sure that the unvaccinated kids are not part of some sort of scale-free network (we'll explain), where epidemics would rage unhindered.

Network topology matters as much as the average behavior of the agents that compose them. Topology matters. If you want to avoid change, or catalyze new change, you need to look at the shape of the underlying networks.

A Network of Laws

When a European parliamentarian introduces a piece of legislation, to be voted on by her colleagues, she may well be making a less independent choice than she thinks. She may only have a vague awareness of the degree of path dependency that her actions are subject to. It appears that the vast web of past laws has a surprising grip on future ones. The past influences the choices she makes today; some recent Greek data archeology reveals how.

Marios Koniaris, with several Greek colleagues, undertook the monumental task of describing and coding the connections between the 50,000 or so European laws that have been enacted over the past sixty years.[3] For the purpose of the study, two laws were considered as connected when one referred to the other.[4] A similar analysis has been made of U.S. Supreme Court decisions, unveiling their own hidden network structure.[5] Koniaris looked at the structures of lawmaking itself for the first time, not just the static body of law. The result was a vast database picturing a huge pile of

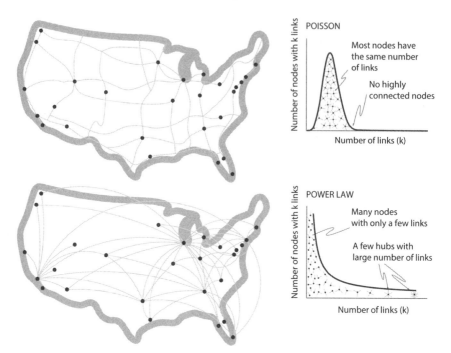

RANDOM AND SCALE-FREE NETWORKS Reformatted from A.-L. Barabási (2015), *Network Science*, image 4.6. http://networksciencebook.com. Accessed May 24, 2019. CC BY-NC 3.0 US.

dour legal text as a giant fishnet, in which each knot is a law and each piece of string represents a connection between laws. With the network thus captured in a database, he and his colleagues could analyze its structure.

The structure of European Union laws obviously wasn't as regular as a fishnet, where each node is connected through a fixed number of connections only to the immediately neighboring nodes. But equally, EU law was far from a randomly connected set of laws. In fact, it is what complexity scientists call a scale-free network.[6] This structure is remarkably pervasive: it is common in our brains, in our circles of friends, or in the architecture of the Internet. The figures above provide a graphic depiction of random networks (above with Poisson link distribution) and scale-free networks (below with power law link distribution). A telltale sign of a scale-free network is the presence of large nodes, with many connections. However, visual inspection is not enough, and a simple plot depicting how often nodes with a specific number of links occur confirms the scale-free character of the network.

Albert-László Barabási is widely credited with the development of modern theory of scale-free networks, but the insight is older. Barabási writes[7]:

"Vilfredo Pareto, a 19th century economist, noticed that in Italy a few wealthy individuals earned most of the money, while the majority of the population earned rather small amounts. He connected this disparity to the observation that incomes follow a power law, representing the first known report of a power-law distribution." Pareto's 80/20 rule is an example of a property of scale-free networks.

And these networks behave in very specific ways; in particular we can know the likelihood where one will grow next. So it becomes possible to assign a probability to which part of the network the next law will appear in, unmasking its path dependence. That is how a network that is invisible or even completely unknown to our parliamentarian was nevertheless gently guiding her political action, unbeknownst to her.

It is also a small-world network.[8] This describes a world that is small in the sense that most nodes can be reached from any node with only a small number of hops. That means that there are big nodes that have judiciously evolved to provide lots of connections, and individual nodes have few enough links to allow them to reach others through these nodes. Plenty of networks are not small-world. Our fishnet is one such example, as it is completely regular, like a taxi dispatch service where everyone is connected to a single node. In the late sixties, Stanley Milgram famously discovered that people were all separated from each other by only six connections. As a result there are only at most six referrals required to connect you, the reader, from the author of this book—another small-world network. This, however, does not apply across time. There are many more links required to map a connection from you, the reader of this book, to Julius Caesar. By adding a time dimension, these social networks lose their scale-free properties. The European body of law, however, did develop over six decades, and that it has nevertheless a small-world structure is remarkable.

We'll pass on the math. But the main point is that networks are not all the same. Their properties really matter to the way they behave. So if we want to change something in the world, it is not enough to refer to it as "just another network." You can leave the math to others, but the various typologies of networks must be understood by policy makers.

Our network of European laws is both small-world and scale-free. Scale-free networks are pervasive and rather amazing. For instance, one interesting property is that this structure makes the European legal code robust.[9] Scale-free networks are efficient to build, as they offer a limited number of connections to get from one point to the next. So every law cites the smallest possible number of other laws, while still being a coherent whole. If our hypothetical parliamentarian had proposed a law that took down another link, it would do little harm to the body of law as a whole. A new law could disconnect a previous relation between existing laws—and thus change the network structure. However, a scale-free network is highly resilient to this kind of random failure, such as making a new law without overseeing all the consequences. This resilience against random failure does come at a price, as it is highly vulnerable to the targeted severance of its key nodes. It is resilient against random cuts of its links, but vulnerable to targeted cuts. Scale-free networks are a type of structure that many natural systems have evolved toward—and even some man-made systems, such as the EU legal network.

Our parliamentarian may well have a populist colleague desiring to undermine the European project and who wishes to purposefully weaken the European legal edifice. Armed with this network knowledge, he should target his legislative arrows at those nodal laws whose obliteration might hasten the collapse of the entire system—and perhaps trigger a revolution.

Similar concerns apply, for example, to the dramatic loss of biodiversity that humans are causing today. The current rate is estimated to be very high: 1,000 times the natural historic rate of extinction.[10] But while few people might mourn the loss of a particular fish, insect, or grass, the real danger lies in the loss of a keystone species. The network of species also forms a scale-free structure,[11] and like the stones in a medieval vault holding up an arch, the keystone species hold together the network of nature. While the gradual extinction is not entirely comforting, the real danger lies in a sudden and highly nonlinear acceleration of loss of resilience of the natural world wrought through the elimination of keystone species. Similarly, a strategically chosen legal amendment could cause system-wide damage in the EU legal framework.

Networks matter, as small interventions can have system-wide effects. To engineer policy revolutions, we need to understand networks.

Sudden Change Is Possible

A nice party trick is to put a couple of Corona beer bottles into the freezer for a few hours, take them out with some care, and then knock them firmly together in front of a few guests. Through the shock, the liquid instantaneously turns into solid ice. Of course, any other beer will work; it's just that Corona is one of the few brands with transparent bottles, so you can see what is going on inside. If this doesn't surprise you, it should. The beer instantaneously switches from a liquid phase to a solid phase through the shock.

Notwithstanding the enormous success of Malcolm Gladwell's *The Tipping Point*, the overwhelming way we think and talk about changes in society is through gradual change. You would be hard pressed to name a tipping point policy. Yet it is the very essence of a revolution that things can become suddenly different.

Phase transitions such as those in a beer bottle are actually quite well understood. Physicists distinguish between various types, but for the purpose of societal transitions, freezing water is a good analogy.[12] What happens in our bottle of beer is that pockets of frozen beer and pockets of liquid beer coexist side by side. Keeping the bottle lying still in the freezer meant that most of these small pockets continued to be independent. Only a sudden shock causes them to connect with each other and suddenly spread the frozen state to the entire liquid.

Just like epidemics spread when small pockets of infected kids are connected through a joint sport activity, suddenly something that was a local phenomenon becomes broadly visible. It is the underlying structure of the network that allows such phase shifts.

The Connected Nature of Innovation

In 1905, Albert Einstein was a frustrated clerk at the Bern patent office. His dream of doing science full-time had been thwarted by the need to support a family and by his lack of sufficient academic credentials. In that year Einstein reviewed dozens of patents regarding the synchronization of clocks. Switzerland has many valleys, and the Swiss are precise. So people wanted to have clocks that had the exact same time in every part of the country. To someone with the imagination and capacity of Einstein, this was painfully boring. Seeking solace in mind games, he started wondering what might happen if you put one of the clocks in motion, on a train. And then make the train go really fast. Like at the speed of light. This thought experiment, combined with the physics that he had mastered, led to the formulation of special relativity theory—reportedly assisted by his physicist first spouse, Mileva Maric.[13] In a way Einstein had not come up with anything really new, but he had brilliantly made new connections between existing ideas. In effect, he had changed the structure of the network of ideas. Thus 1905 became known as the miracle year in physics. Einstein published at least four such radical scientific breakthroughs in a single year, working in his spare time. It is the very lack of resources that shows how he was brilliantly recombining and changing the structure of the network that connects the ideas of physics.

Innovation is best understood as a network where new nodes grow from existing ones by making new connections. Brian Arthur is an economist and one of the very early complexity scientists at Santa Fe Institute. In a delightful short book, *The Nature of Technology*, he reflects on its nature. He describes eloquently that contrary to popular belief, new ideas do not fall out of the sky, but, like relativity theory, they are formed from the inspired recombination of existing ideas.[14] In this innovation network the nodes are existing technologies, which recombine to form new nodes.

Network Economics

Characterizing networks as different as European law, freezing beer, innovation, and vaccination with a single language and tools leads to a better insight into these problems. Network analysis has also taken a toehold in economics.

When you choose a brand of smartphone, you'll likely look at what brands your friends have. Even if there were a perfect consumer report, allowing you to make a perfectly rational economic decision, most people can't be bothered. So your decision depends on the decisions that others have made before you. Your decision is path dependent. How many friends you have also influences your choice. Economists call this the network effect. Economic agents are not independently optimizing a utility function, but their decisions depend on others' decisions.

But what if one of your friends is the opinion maker for the group? Or if there are several opinion leaders, connected with clusters of followers? Or if some friends have different numbers of friends themselves? You can see that the structure of the network can rapidly become more and more complex. That complexity will result in different smartphone buying decisions. For an individual decision this is all far too complicated, but when considering policy changes, the nature of the network matters a lot.

Take corporate taxes.[15] Corporations put considerable effort into routing their profits through countries so that taxation is lowest. This is known as treaty shopping. For example, profits transferred from the United States to the Netherlands may be taxed at 20 percent on arrival. But sending the same funds to India results in a 10 percent tax on destination. Sending them on to the Netherlands from India is no longer taxed on arrival. So routing profits from the United States to the Netherlands via India results in halving the tax bill. The rates in the tax treaties between the main 108 countries form a network. Once the network is modeled and mapped, policy makers and treaty shoppers can study its properties and intervene. The UK, Luxemburg, Estonia, and the Netherlands are the nodes in this network through which most money is routed. Companies cut their taxes roughly in half by navigating this network. Only when it is mapped and understood can countries devise effective policies to either attract larger flows or endeavor to stop tax avoidance—whichever happens to be their priority.

A Divestment-Driven Revolution?

Rockefeller money is oil money. It was amassed during the first half of the twentieth century, as the world started to power itself using internal combustion engines. In a shock and irony move, in the summer of 2017 the Rockefeller Brothers Fund announced that it was divesting from fossil fuels.[16] Activists and pundits cheered the move as a significant blow to the industry. But was it? Notwithstanding its wealth, the fund owned only an infinitesimal part of the industry. Also, the shares it sold were simply bought by someone else who was perfectly happy to hold them. Commentators said that it was an important precedent and signal, the start of a revolution with institutional investors. But it is significant only if the signal is followed and spreads widely.

In order to understand whether an event such as the Rockefeller divestment matters, we need to understand the network that might allow the event to propagate and lead to nonlinear change. Different networks behave quite differently, and as a consequence you need to know about types of networks. Only then can you start to assess whether an event can plausibly start a revolution—or whether it is simply an isolated anecdote.

How can we know whether something like the Rockefeller divestment is just a notable but isolated event, or the start of a significant change? Ideally we'd want to map the network structure between investors to see whether they perhaps form some kind of a scale-free network and what type of contagion exists across its links.[17] We could then model its behavior and conclude from that. Except realistically we don't have all that information—so in world with imperfect data what can we say?

In the past people have similarly tried to influence events by encouraging divestment from the tobacco industry and from South Africa over apartheid. It typically starts with a few faith-based or public groups very visibly divesting and taking a stance, but the amounts involved are small. This subsequently spreads to pension funds and university endowments that then sell their holdings.[18] That is quite a good start—but still only a drop in the bucket, as the financial markets are so large.[19] And for everyone who divests a stock, there is someone else who buys it. Only when there are many more sellers than buyers will the stock fall and the divestment campaign have achieved what it set out to do.

The real question is what the contagion mechanism might be to go beyond the initial divestors. This is the hard work of building connections between groups that do not normally listen to each other, about creating a shift in value leading to stigmatization of the incumbents, ultimately edging the system toward the point of phase transition after sufficient contagion steps.

For divestment campaigns historical precedents are not encouraging. Not a single campaign has been successful through the financial markets. In some cases they have been instrumental in changing opinion and supporting other legislation. This was the case for extensive boycott measures against the South African apartheid regime, for example.

So was Rockefeller exiting fossil fuel investments really more than anecdotal gossip? Barely, except if the reporting had been coupled with a discussion of what it takes to create a contagion mechanism to get to scale, along with a discussion on how big the fossil fuel system really is. But such systemic journalism is rare. When you are looking for a revolution, individual actions are important, but without a sense of the underlying network, they remain only individual actions. Even if you are the Rockefeller Brothers Fund.

Networks Literacy

Being network literate helps you understand the trade-offs in policy better, even without being able to quantitatively model the network, as the Greek data archeologists did for EU legislation. Network science is very useful when you can apply it through modeling—but it already helps a lot just to know about its concepts to frame a problem more powerfully.

Basic network literacy offers an understanding of how sudden change might happen—the stuff of revolutions. The first step is to realize that this kind of change is possible and that it doesn't require a huge input to get a huge change. Call it disproportionality. Small actions such as suppressing for-sale signs can help avoid segregation. Curbing mercury emissions can help tip an entire fleet of coal plants into early retirement, even outlasting the government that put the policy in place. In Chapter 8 we'll see how network effects are essential to understand and influence the evolution of social norms. And influencing norms is a core ingredient of climate policy.

$$\textcircled{4}$$

Non-equilibrium *Is* the Source of Order

In fact, if the system ever does reach equilibrium, it isn't just stable. It's dead.

JOHN HOLLAND

Equilibrium is a familiar idea. A child on a swing uses her weight and balance to move away from the equilibrium point where the swing hangs still in the middle. Sustained physical efforts must be exerted to keep it from reverting inexorably to the point of equilibrium, the swing hanging limply under the bar from which it is suspended. Equilibrium is an idea that is still central to much of science, prominently in physics and, more relevant for our purpose, also in economics—notably in the elaborate models that are used to evaluate climate policy. Equilibrium in economics leads it to play a foundational—and largely implicit—role in policy considerations.[1] Various scientific disciplines distinguish between many different kinds of equilibria, but for our purposes the intuitive idea is straightforward enough.

For a long time scientists thought that equilibrium and temporary deviations from it—for example, for a swing—were a reasonable way of describing the world. In this view the world is in balance, in a stable state of being. In 1984, taking a radically different view in a sweeping treatise on the subject, Nobelist Ilya Prigogine articulates the cumulative scientific insights from multiple disciplines as follows: "At all levels . . . nonequilibrium is the source of order. Nonequilibrium brings order out of chaos."[2] Prigogine frames reality as in a constant state of becoming, rushing from one unstable state to another, constantly being created. These are not the loose ruminations of an amateur philosopher, but the considered reasoning of a renowned natural scientist, digesting the cumulated insights of a century of fundamental breakthroughs in our understanding of how the world actually works. In contrast, economists—and through their influence policy makers—have firmly continued to embrace a view of society as either being roughly in equilibrium or in temporary deviations from it. This happened in harmony with other sciences in the latter half of the nineteenth century. The familiar charts balancing supply and demand through price are an illustration of that perspective. And while for many societal issues, this is not unreasonable, and the equilibrium perspective is helpful, it necessarily excludes the possibility sudden change—and therefore of revolution policy. Hence our interest here for this seemingly somewhat philosophical disconnect—but one with a profound practical consequence.

Bringing up the first and second laws of thermodynamics at a dinner is a guaranteed conversation stopper. But it shouldn't be. They are as fundamental to how we understand the day-to-day world around us as cooking a meal or politics. The first law simply states that energy is conserved. There are a few details. For example, energy can change from one form to another, such as heat into motion when you burn fuel in a car engine. Or when you use electric power to generate cold in your refrigerator. The other critical detail is that this holds for isolated, closed systems. The law was formulated in the middle of the nineteenth century, and this timing played a crucial role in the history of economics, as we will see.

The second law states that any closed system tends to a greater state of disorder. The formal measure for order is entropy, and it always increases. A familiar version is the tendency of a child's room to fall into greater disorder, as toys and clothes inexorably spread. It requires the addition of some effort to put everything back into closets, drawers, and bins. What holds for a child's room holds for the universe. However, the consequences of the second law are vast, and although it was first formulated in the first half of the nineteenth century, its full impact was not deeply understood until more than a century later, notably by Prigogine. Suffice it to point out here that it introduces a direction of time; the future is different from the past.[3] This is not absolute, but probabilistic. Martin Gardner writes in *The New Ambidextrous Universe* that "certain events go only one way not because they can't go the other way, but because it is extremely unlikely that they go backward."[4] The fundamental asymmetry between the past and the future is generative for changing any system. It is what allows systems *to become*, instead of just *to be*.

How Economic Modeling Came to Embrace Equilibrium

Léon Walras was born in 1834, the son of a school administrator. He studied engineering at the still prestigious École des Mines in Paris, but soon grew tired of it. Abandoning his studies to indulge his bohemian temperament, he turned to journalism, even producing several unsuccessful novels before, at the age of twenty-four, his father coaxed him to focus on economics. Earning a living with various banking jobs had given him some appreciation of the real economy, but the only formal social sciences training he received was from his father. He correlated those insights with the basic engineering knowledge he had picked up during his stint at the École des Mines. Denied a professorship in his native France, Walras was offered one in Lausanne in 1870, at the age of thirty-six. He did achieve recognition during his life, and his ideas had a very substantive influence. In 1892 the American Economic Association elected him an honorary member in recognition of his "eminent services to the Science of Political Economy." But his main work, *Elements of Pure Economics*, was not translated into English until 1954.[5] His influence was indirect, as he became the first exponent of the increasing mathematization of economics that played out in the first half of the twentieth century. Economics endeavored to leave the orbit of the social sciences, where it had been born. In hindsight this was an oversimplified mathematics, one made tractable notably through the assumption of equilibrium. It was an artifice that made the formula's solvable.

Walras's grand insight was to characterize economics as a mathematical model, taking inspiration from his engineering training. He chose an equilibrium model, based on the first law of thermodynamics. As we mentioned above, although the second law had been formulated, its profound impact on the ability of a system to evolve and change was not appreciated until many decades later. And most certainly Walras did not grasp its foundational impact, and he ignored it, locking in macroeconomic modeling for the foreseeable future.

Eric Beinhocker has taken an unusual journey from his perch as a McKinsey consultant to become the head of the Institute of New Economic Thinking at the University of Oxford. In his book *The Origin of Wealth*, he describes the history of equilibrium thinking in economics in great and erudite detail.[6] He underlines its centrality, quoting the former chief economist at the International Monetary Fund, Olivier Blanchard: "Macroeconomics today is solidly grounded in a general equilibrium structure. Modern models characterize the economy as being in temporary equilibrium, given the implications of the past and the anticipations of the future."[7]

Obviously, many economists have engaged with these arguments, and many are conscious of the limitations of this predilection for equilibrium.[8] And while there are ongoing efforts at developing theories of economic complexity, the discipline has been loath to abandon the core idea of equilibrium for its big models of the future. This in turn has had a defining impact on climate policy. The response has been to consider temporary deviations from equilibrium and the inclusion of externalities, rather than moving to non-equilibrium models.[9] In fact, the very idea of an externality betrays how you conceive of the system. Take environmental factors. The standard approach is to view them as an externality to the economy and to reflect their cost by adding it into the economic models. Contrast that with a view that sees that nature and society are deeply intertwined and influencing each other in a myriad of ways and many scales, some understood and others overlooked—in essence a two-way relationship and not something to add in. For all sorts of short-term decisions the externality approach may well be a reasonable approximation. But it becomes counterproductive for long-term systemic issues such as climate policy.[10] So while the assumption of equilibrium may be approximately right, it is in principle wrong. Better to think of today's economy as a temporary snapshot in a state of non-equilibrium, rather than the other way around. In any case, the implication is that for our purpose of understanding how to foster revolutions through policy, we must tread outside the standard frame of equilibrium economics.

The High Cost of Free Parking

Part of what makes Cinque Terre on the Italian coast such an amazing tourist destination is the total absence of cars. The streets are too narrow, the terrain too steep. An important attraction is the hike along the coastal path from one village to another. Or so goes the lore. In fact, there are websites, mostly those catering to driving Americans, that assure concerned would-be tourists that every village and town in Cinque Terre is accessible by car. More importantly, those write-ups don't just discuss getting there, they talk about what to do with your car once you arrive. Yellow markings on the street are for residents, blue for visitors. Though it's best to avoid both and hunt for white lines, which mean *free* parking.

Of course, those spots are anything but free. More likely than not, they are full. So unless you'd like to arrive before anyone else wakes up, or drive when no one else does—and, thus, demand for parking is low—the "free" spots will be taken. Otherwise, the price for nominally free parking will mostly be paid in time—circling around in search of parking—or perhaps by thwarted attempts to visit any one particular village in Cinque Terre altogether: "No parking here, let's keep driving."

If the cost of switching one's vacation plans to accommodate the need to park the car isn't high enough, there are plenty of other costs that come with free parking. *The High Cost of Free Parking*—a book with an explicit name—details a whole host of instances where parking is anything but free: to individual drivers, to those living around them, to building owners, to cities, to society, and to the planet.[11] Everyone else pays when those enjoying any one particular good or service don't.

Privatizing benefits while socializing costs is hardly ever a good idea. It motivates those who benefit from any one particular action to do too much. Subsidy is another word for it—and not the good kind of subsidy that rewards learning by doing and that leads to deliberate progress down the road. It's simply a subsidy that encourages too much of a bad thing.

That's not to say that driving is bad per se. Too much of it is, though. And subsidies for drivers abound. It begins with free roads. Roads, of course, aren't truly free. The government typically pays for them. And indeed, U.S. highways, much like the German Autobahn or local access roads, provide a real public service. Mobility is a good to be cherished. It connects. It makes us more productive. But all of that also comes at a cost.

Pricing Works . . .

The costs of *free* are perhaps nowhere as obvious as with parking. Cinque Terre, too, might have free parking spots, but at least the towns themselves weren't built for cars. The typical town square is just that: a place for townspeople—and tourists—to gather, sip macchiatos, and gossip. Cinque Terre's villages might be accessible by car after all, but what makes the villages themselves such a livable place for some and a magnet for many tourists is that their inherent natural beauty, not the use of cars, defines them.

Think of cities you'd call well designed, and you probably picture similar arrangements. The most desirable neighborhoods are those that put people first: Jane Jacobs made these arguments in the 1960s, in a now classic book entitled *The Death and Life of Great American Cities*. She describes the city as a complex system that has grown and evolved around and through people: "This order is all composed of movement and change, and although it is life, not art, we may fancifully call it the art form of the city and liken it to the dance—not to a simple-minded precision dance with everyone kicking up at the same time, twirling in unison and bowing off en masse, but to an intricate ballet in which the individual dancers and ensembles all have distinctive parts which miraculously reinforce each other and compose an orderly whole."[12]

Desirable also means costly. Or we should say that the fact that these kinds of walkable neighborhoods command outsized real estate prices implies that demand for them outstrips supply, almost everywhere you go. The higher price ensures that demand and supply remain in balance. It's the Law of Demand: price goes up, quantity demanded goes down.[13]

Now think of cities or neighborhoods that aren't quite as desirable to live in. The word "affordable" might go hand in hand with them. Real estate prices will likely be lower. That lower price reflects one of two things: either there are more of these types of cities out there relative to demand for them or, conversely, demand for these kinds of cities is low relative to their supply.

Proper prices fulfill many important roles in our economy and society. They are a central organizing device.[14] They allow the economy to function— to find the *right* amount of a particular good or service being bought and sold—without anyone calling the shots. That's the fundamental principle of a market economy, but it only works under certain conditions. The primary one is that prices reflect the true cost of a particular good.

Every mile driven engenders certain costs. The primary costs to the driver are the car itself and the fuel needed to drive it. There's also time, in particular the opportunity cost of time spent behind the wheel. Then there are accidents. In fact, accidents alone amount to around $2–$3 per mile driven.[15] Drivers themselves don't pay these full costs. Others do. Society does.

Setting the Right Price

The traditional economic formula, then, is to internalize all externalities, to make everyone pay for the true cost of the good or service they enjoy. At the COP21 climate conference in Paris, Elon Musk was to give a much-anticipated speech to the 1,000 or so assembled mayors at the Hotel de Ville. He climbed the platform, adjusted the microphone, said "Price carbon," and left.[16] In the climate context, the simplest such economic pronouncement comes down to a carbon tax. Each ton of CO_2 emitted today causes significant damage over the many centuries it stays in the atmosphere. Each ton emitted today, then, also has one *right* price. Make everyone pay that price, and the *right* amount of CO_2 will be emitted.

For driving, that implies that each gallon of gasoline, each mile driven, will now be more expensive—and appropriately so.

And don't stop there. Carbon emissions are only one of a huge number of negative externalities associated with cars. Don't just price the carbon externality; price other more local pollution externalities, price the accident externality, the traffic congestion externality, the parking externality. In short, set the right price, and get out of the way.

That would indeed be a good start, if it were that easy. For one, pronouncements like that ignore political reality. It takes cheaper, alternative technologies to make pricing the offending technology a viable policy intervention in the first place. For climate policy, that often means pushing alternative technologies much more directly.[17] But the complications don't end there.

Pricing the Network

Not all prices are neatly determined by the interplay of demand and supply in a competitive market. That's a slight understatement. Very few, if any, prices follow that ideal. Prices for any one particular good aren't set in a vacuum. Even standard markets aren't always competitive. Monopolies, duopolies, and oligopolies indicate various forms of market concentration on the side of suppliers. Monopsonies, duopsonies, and oligopsonies are their equivalent on the buyers' side.

Less familiar but more important: there is no obvious or automatic way to set a price for many things of value in complex systems. What is the price of an antibiotic in view of the exponential increase in bacteria that are increasingly resistant to all known formulas? If today you take an antibiotic, you are arguably putting someone else at risk. Your taking of the drug will slightly increase the resistance of bacteria to it, exposing someone in the future to a disease that is less curable. The Easter Islands are barren, with no trees. In what is assumed to be an early environmental debacle that doomed a civilization, the loss of trees meant that the inhabitants no longer had the means to sustain their livelihood. In this perspective, how to you price a tree when there are only a thousand left? How to you price it when there is only one left? When everything is connected to everything, you can't compute a price that reflects the real value. Considering it an externality only prices in the immediate—first-order—effect, not the full system impact.

Pricing is often framed as the visible sign of the invisible hand at work, balancing the economy. However, here we consider pricing from a policy perspective, as a tool for an "activist laissez faire" policy and for shifting the system to a new equilibrium, overcoming path dependency.[18]

Consider price as an equilibrium point shifter: The usual approach of gradually increasing the cost of carbon works under the critical assumption that the end state can be reached through incremental change. If that is wrong (as we believe), there is a real danger of the boiling frog syndrome. We adjust to the price pain of lack of ready alternatives; by the time the alternatives are readily available we are no longer able to adjust. The time delay has strengthened the lock-in. An alternative approach is to start with a high carbon cost to shift the system into a different equilibrium, combined with active, strong subsidies for innovation.[19] Soon after the initial jolt, the carbon cost can come down in the new normal.

Revolution Needs Complexity

Complexity does not necessarily mean more moving parts, or more of the "representative agents" that macroeconomic models use. What matters is capturing the most pertinent details for modeling a particular set of behaviors. In the end, it is often a phase change that dominates the final outcome. This means taking in the full lessons of the second law of thermodynamics. Any equilibrium is at best temporary or, more accurately, an approximation that helps us make sense of things. But it is approximately right and in principle wrong, particularly if we are interested in change. It means recognizing non-equilibrium as the root of the capacity to change. As John Holland is quoted as saying in the epigraph to this chapter, a system in equilibrium is dead. It means seeing society in a constant state of non-equilibrium, becoming and transforming at all times. Extensive path dependencies provide anchors of stability and create the illusion of predictability.

Our grand ambition is to describe a way of framing revolutionary change, in order to bring change about more purposefully, at the right time and the right place. Carbon pricing seems like good idea, but it assumes an equilibrium system that can be efficiently changed. So even if we could agree to implement such a tax, its impact will likely fall short of expectations.[20] It has become an article of faith among climate economists that carbon pricing is the cornerstone of any successful climate policy. An ever-louder choir ranging from Exxon to Bono have echoed this point. Except that both in practice and in theory there is very little evidence that it actually works. In those geographies where it has been implemented, such as California or the EU, prices have been too low to make any substantial difference. In the EU, emissions fell slightly, but the carbon price fell even more. In 2017 emissions covered by the carbon trading scheme even rose. It remains a good idea, at least in principle, as it focuses attention, raises money, and provides some price signal for climate action. But any politically acceptable price will be too low to shift the system to a different state; in practice it works only at the margin, as pricing theory would predict. From a complex systems perspective it becomes obvious that a price high enough to lead to phase transition is unrealistic, and the realistic price only has a marginal effect cementing the status quo. In Jeffrey Ball's words, "The result is that a policy prescription widely billed as a panacea is acting as a narcotic. It's giving politicians and the public the warm feeling that they're fighting climate change even as the problem continues to grow."[21] By all means price carbon, but it cannot be the cornerstone of climate policy.

Pricing carbon is a good thing, but only if the underlying network is ready to be triggered and the political economy allows you to set a price that can trigger a phase transition. Do it, if you can—just don't put all your eggs in that particular basket. Going beyond equilibrium systems is a crucial first step. One way to do that is framing today's societal system as a temporary snapshot in a state of non-equilibrium, rather than the other way around. It is the key to unlocking a future different from the past—to have it *become* instead of merely *being*.

Top-Down and Bottom-Up Order

Unfortunately, many analysts—in academia, special-interest groups, governments, and the press—still presume that common-pool problems are all dilemmas in which the participants themselves cannot avoid producing suboptimal results, and in some cases disastrous results.

ELINOR OSTROM

Graz is an orderly and picturesque Austrian town. But head to the Sonnenfelzplatz and witness a seemingly spontaneous waltz of buses, bicycles, pedestrians, and cars careening around each other, without any semblance of order. There are no traffic lights, no sidewalks, no bicycle paths, no bus lanes, no lines on the road, and no traffic police. Yet the traffic smoothly weaves.

Traditionally, traffic is a bastion of the principle of top-down order. Little is left to chance as flows are micromanaged through a proliferation of rules and signs. Poor outcomes in terms of widespread injuries and rampant congestion are excused by a lack of alternatives. Enter Hans Monderman, a Dutch traffic engineer, who coined the idea of "shared space." His conviction was that by getting rid of every inch of micromanagement, traffic would be safer and flow faster. His first accomplishment was convincing city authorities to try his idea. Imagine a credulous mayor facing the angry citizenry if the experiment had claimed a traffic victim. How could she explain that this way of organizing a system had been a reasonable thing to try? Yet it is reasonable in the right context. There is a delightful suite of videos that show Monderman walking backwards across his first shared space intersection in the northern Dutch town of Drachten, all the while being interviewed. The idea has since caught on in several Dutch cities and beyond, as in Graz.

A casual observer will conclude that this is pure bottom-up organization at work. One with a libertarian bent might add that it demonstrates that fewer rules lead to better performance and that the state should shrink further and leave space for the market and bottom-up order. A more contrarian observer would say that this doesn't prove anything and that the traffic chaos in Lagos or Calcutta demonstrates that traffic must be coerced with an iron fist. An observer with some knowledge of complex systems would state that this precisely shows that the source of order is neither bottom-up nor top-down, but stems from the relationship between them.

There is no shortage of rules in the Netherlands that impact traffic: primary schools have compulsory traffic education, drivers need licenses, cars undergo regular safety inspections, and courts deliver timely penalties for any transgression. Additionally, Dutch society has a relatively high trust culture. This enables the right amount of give-and-take to allow others to pass, and to be considerate of the needs of others. But it required purposeful state action to decide to make shared space a design principle for traffic intersections, to fund the changes in physical infrastructure and to put in place and to maintain the myriad rules that facilitate traffic. All these actions and the context of social norms that had evolved over time enabled traffic to self-organize and deliver an effective and superior bottom-up order. This is nothing less that the purposeful design of an ecosystem.

Shared space is not a panacea. It has proven to be effective in northern European regulatory environments and cultures. Graz vice-mayor Lisa Rücker downplayed objections to the project, stating, "I assume if the Dutch can do this, if the Germans can do this, that also the Austrians can do this."[1] A harder question is whether it might work in Greece or in Egypt. Or whether additional targeted measures would allow the system to succeed.

Self-organization is a lot easier when people are honest and trustworthy. This makes the transactions easier and faster—and reduces the need for an external authority such as a traffic light. In a particularly devious research project, Alain Cohn and his team distributed 17,000 wallets with and without money on the streets and in the cafés of 355 cities worldwide. Each wallet identified its owner with a simple card. Strikingly, in all countries a wallet with money was more likely to be returned than one without. But the percentage of wallets that were returned ranged from the high seventies in Norway and the Netherlands to the low twenties in Kazakhstan and Kenya. The root causes are manifold; contrary to what you might expect from economic incentives, people were more likely to return the wallet when it contained more money. It is notable that all the countries where shared space has been successfully applied to let traffic self-organize are in the top quartile of wallet returners.[2] Honesty and trust ease the process of self-organization, as it makes the transactions between the agents in the system easier.

To some extent these questions can be explored by using complex systems models, but the success of shared space is deeply connected to an understanding of human behavior. And this is hard—but not impossible—to model. Our interest here is not to solve traffic congestion, but to gain insight into the merits of top-down or bottom-up action to deal with the climate crisis. Unfortunately, considerable confusion has been sown on these organizing principles by the two narratives that represent this dichotomy in our politics.

The Market vs. State Dichotomy

Traffic is a caricature of the entrenched tussle in Western politics between the primacy of the state or of the market. In Ronald Reagan's 1981 inaugural address, he famously said, "In this present crisis, government is not the solution to our problem; government is the problem." In contrast, philosopher Noam Chomsky acidly notes, "Haitian rice farmers are quite efficient, but they can't compete with U.S. agribusiness that relies on a huge government subsidy, thanks to Ronald Reagan's free market enthusiasms." Contrasting positions on the primacy of the state or the market will be familiar from many Thanksgiving dinners, newspaper editorials, and political debates.

In an earlier book, *Complexity and the Art of Public Policy*, authored with David Colander,[3] we described how this contrast has its roots in one of the most curious chapters of intellectual history. We tell the story of how during the course of the twentieth century the discipline of economics bifurcated into competing schools of thought that advocated either for the primacy of the market or of the state. This in turn enshrined a standard policy framework in which the two are irrevocably opposed. In fact, that book came about as a result of our meeting at a climate policy conference in Berlin, where a debate raged between those who advocated state or market solutions to the climate crisis. So strongly separated are these positions that a fair proportion of the climate change denial among U.S. Republican politicians may be attributed to a belief that climate policy necessarily implies robust state interventions. Rather than consider these, better reject climate change in the first place. Some, such as the Heartland Institute, a U.S. think tank, have even made this argument explicit by contending that a stronger role of the state does greater damage to economic health and social freedom than any devastation that climate change might unleash.[4]

The standard policy framework mixes the goals and the means: a free market or strong state action are the means to accomplish something, for example, goals such as greater freedom or solidarity, a cleaner environment, an accepted level of inequality, or religious tolerance. Strongly diverging views on those aspirations are healthy ideological differences. Strongly diverging views on which tools to use just suggests confusion. The tools should be the subject of a more technical debate, rather than a political one.

In a complexity frame, state action and market dynamics become mere tools to implement policy goals. They are not the goals themselves. To address the climate crisis, we must design a judicial mix of government action and market mechanisms, and not get stuck in a sterile argument about whether one is better than the other.

An additional perspective that is illustrated by the Graz traffic story is that the source of order in a system is neither pure top-down from government action, nor bottom-up from market dynamics. The magic of efficient self-organization occurs when the relation between the two forces is understood and properly balanced. In *Complexity and the Art of Public Policy* we characterize this as "ecostructure policy." This is the purposeful top-down design of a structure that allows for bottom-up dynamics to play out.

The idea of ecostructure policy also requires reconsidering the place of government.

Government from Within

When you're growing rice on terraces, you have to make sure that the farmer above you doesn't throttle your water supply, lest your fields go dry and your crops wither. That in a nutshell is the governance problem of rice terrace systems. In the standard policy lens, this suggests the need for a strong government to impose rules for water allocation, or for a market system to allocate water efficiently.

But there may be a better way.

In 2012 the *subak* system in Bali was designated a UNESCO World Heritage Cultural Landscape, described as follows:

> The cultural landscape of Bali consists of five rice terraces and their water temples that cover 19,500 ha. The temples are the focus of a cooperative water management system of canals and weirs, known as *subak*, that dates back to the 9th century. Included in the landscape is the 18th-century Royal Water Temple of Pura Taman Ayun, the largest and most impressive architectural edifice of its type on the island. The *subak* reflects the philosophical concept of *Tri Hita Karana*, which brings together the realms of the spirit, the human world and nature. This philosophy was born of the cultural exchange between Bali and India over the past 2,000 years and has shaped the landscape of Bali. The *subak* system of democratic and egalitarian farming practices has enabled the Balinese to become the most prolific rice growers in the archipelago despite the challenge of supporting a dense population.[5]

Farmers meet regularly at the water temples to discuss water management. Neither an external state nor a market governs their action; rather, it is the combination of the sacred nature of the temple with the practical need to allocate water resources that led to a sustainable and successful governance system.

Stephen Lansing is an anthropologist and complexity scientist who has spent much of his career studying and documenting the Balinese system of rice production and governance. Together with other complexity scientists, he shows how the governance of the system is not external, but an emergent property of the system itself. The combination of physical characteristics of the land with the goal of cultivating rice with a group of independent farmers leads to experiments in governance. These subsequently recombine and evolve over time to the current structure. A computer simulation for the *subak* system shows how the governance grows, as it were, from the system itself. It is inside the system, not superimposed. Remarkably, those simulations also illustrate why attempts to increase yields in the Subak through Green Revolution methods led to chaos in irrigation and devastating losses from pests: government from within.[6]

Growing rice, even doing so successfully over ten centuries, is a pretty simple problem compared with governing a modern technological multicultural society, let alone tweaking it to deal with the climate crisis. But it illustrates clearly how the government is an integral part of a complex system, connected to it, coevolving with it, and even emerging from it. The core contrast between the complexity frame and the standard policy frame is that the government should be seen as being inside the system. It is not an external force that somehow acts independently on the system.

One of the major hurdles for effective climate policy is the failure to reconnect with this idea of government from within. Report after report is published, recommending to governments and businesses what actions they should take to fix the climate. But this approach assumes that governance is somehow largely independent of the system and is able to take action based on analysis. The reality is more, well, complex. If we consider government as inside the system, it shifts the problem and requires difference solutions.

That we persevere in this approach is surprising, as it runs counter to decades of insights from social science.[7] This will tell you everything you need to know about interest, alignment, incentives, and the importance of institutions. Social scientists understood complexity well before complexity scientists. With the exception of many economists, they do not assume equilibrium. Max Weber, who is considered one of the founders of sociology, described the benefits of a bureaucratic state as enabling the primacy of knowledge over power because it is rooted in society. Karl Dusza noted in a

1989 article that "Weber was fully aware that combined action of a plurality of individuals results in something *sui generis*, which cannot be reduced to the action of the one, not even of the multiple ones. From the combination of 'unit acts' there emerge ever more complicated structures, the complexity of which at one point reaches a degree where they appear as an objective, external bond standing over against the acting agents."[8] Government is described as emerging from the societal system and intertwined with it.

So what can complexity even add to the debate? While many social scientists covet the policy prominence granted to economists, most are unwilling or unable to face up to the formal modeling and statistical structured arguments that have given economics its prominence. Complexity has the potential to provide social science with both a shared language and a common framework that can strengthen the impact of its contributions. In a reflection on how complexity might reintegrate the social sciences, we described the challenge as follows:

> Adopting this common framework will require significant change in the thinking of the various sub-disciplines of social science. For example, economists will have to recognize that the underlying social theoretical model needs to be much broader than the one they have recently used. It must be able to incorporate sociological, cultural and political insights, and that the problems faced by society are much more complex than their current standard atomistic models recognize. Sociologists, on the other hand, will have to recognize that the development of a rigorous social model is necessary for scientific advancement, and that such a model need not be designed to rule out their insights, but simply to formalize them in a way that will allow modern analytic tools to integrate those insights with insights from other fields.[9]

Considerable progress is being made in this direction, even leading some to coin an entirely new discipline of computational socioeconomics.[10] They make this claim in what is no less than a manifesto of a new interdisciplinary field of research. By combining data with theoretical models in multiple disciplines, new insights have been gained for problems ranging from inequality to industrial structure, economic growth, the monitoring of emergent events—or even the naming of babies (hint: it is subject to clear and quantifiable network dynamics).

Three Models of Governance

Complex systems are not in equilibrium. In fact, as we saw, they are best described as being far from equilibrium. What does the governance of non-equilibrium systems look like?

The end of a life is a real revolution; one state irrevocably replaces another. Brian Walker is an eminent complexity scientist from Australia who describes some of the dynamics involved in an accident.[11] He relates the experience of a friend, an emergency room (ER) specialist, who characterized a patient after a serious accident as teetering on the crest between two basins—a basin of life and a basin of death. The first thing that doctors do upon admission of a seriously injured patient is to attempt to stabilize vital statistics such as heartbeat, blood pressure, and so on. In a sense this settles the person on the crest between the two basins. The next challenge is to allow the body's system to self-organize itself to glide back into the life basin and not into the death basin. Brian's friend, the ER doctor, noticed a curious phenomenon. If the doctors managed the vital statistics too tightly, this got in the way of recovery. By giving enough space for the parameters to bounce around, the system was better able to find its way back to the life basin. It needed, as it were, to be able to explore the full parameter space in order to enable it to self-organize and survive.

Similarly, for revolutions to be able to coalesce, the strings must be loosened enough to let the system play and learn its way to a phase transition. Fortunately, most policy choices are not as dramatic and stark as the balance between life and death for an individual, although some undoubtedly impact the life and death of a great many people. Such as the way the single act of a Gavrilo Princip can tip the world into war. Or when Facebook's early design choices inadvertently allowed for election meddling and putting someone's finger on the scales of democracy. Policy has consequences. The point is not to be dramatic, but to introduce the image of a system with two basins as a helpful lens. Think of the current state as a pebble rolling around. An equilibrium system would be a single basin, where any temporary departure from the bottom would over time lead the system back to its original equilibrium state. Now imagine that the goal of the policy maker is to get the pebble from one basin into another. One basin could represent the current transport system, and the other basin the future mobility system of autonomous vehicles. And critically, it would also include all the associated changes we have discussed, in terms of energy use, urban design, social norms, and so on. What is the most plausible way to get from one basin to another? We distinguish three options:

You can use force. An external force commands the pebble to move out of one basin into the other. This requires a couple of things: sufficient authority, and the conviction that the path is knowable and that the given order will actually take it over the rim into the next basin. We'll call this the top-down approach. Alternatively, you can take the view that with the appropriate nudges, stakeholder engagement, and proper analysis, market forces will take the pebble to the other basin. This is the incremental approach. The big question is whether this will provide enough momentum to take the pebble over the edge into the next basin, or whether the pebble will merely end up rummaging around in the first one. Finally, using all the approaches we have described in Chapter 4, one can attempt to change the very dynamic of the system, in the expectation that it can rejig itself enough to coax the pebble over the edge. That would require enabling the kind of phase transition, the sudden reconfiguration of the underlying networks, that we have described in earlier chapters.

The distinction between these three approaches is not an exclusive framework. They form more of a continuum, but at their core they have different organizing principles.

1. Top-Down: The Familiar Top-Down Planning Process

A more or less centralized source of authority makes decisions and issues instructions to implement them. The system is framed as never far from equilibrium; representative agents are central to model policies, and path dependencies are of secondary importance.

The narrative about the system matters, as well as the reality. Even in states that are described as authoritarian, the reality is more nuanced. In highly centralized Singapore, I asked some of my students to draw a chart of the decision-making and influencing actors for elderly care. A rich and complex web of civil society organizations, businesses, and government departments appeared on the whiteboard, with varying degrees of interconnection and influence. Yet even policy-savvy Singaporeans refer to their nanny state being all-powerful and having its fingers in everything. But that is an oversimplification. Top-down planning is not a caricature of a reductionist world where everything is known and plannable, but a broad policy approach where decisions are mainly described as made by a central set of actors, and underpinned by forecasts, plans, cost-benefit analyses, and trade-offs.

2. Incrementalism: The Iteratively Improved Minimum Viable Policy

Laudably inclusive and thorough, an iteratively improved minimum viable policy represents a different approach from a top-down planning process, as it involves a broad swath of stakeholders. But it does not aim at rocking the boat. The end point is envisaged to evolve gently from the current state, protecting the interests of incumbents as best as possible, through negotiated trade-offs.

No reduction in CO_2 emissions since 2009. Zero. "The Kingdom of Orange is not very green," journalist Ben Coates quipped about the state of environmentalism in the Netherlands.[12] After 225 days of negotiations, in 2017 a new Dutch center-right coalition government was unveiled on the steps of the Royal Palace after having been sworn in by the King. The surprise was that a commitment to radically reduce greenhouse gas emissions had been inserted into its plans. In fact, it aimed for the country to leapfrog to becoming a frontrunner.[13] Planners did their sums and allocated reductions to five sectors in society; over 500 organizations, companies, and coalitions would meet weekly over a year to hammer out incremental steps toward the final goal.[14] The outcome, predictably, was a plan that fell short. Evolution may well be the only choice under the circumstances, but it is not revolution. The implicit goal is tweaking the status quo, not transformation.

3. Revolution: The Purposefully Disruptive Policy

This is where all the elements outlined in Chapter 4 come into play. We deal with path dependencies and lock-ins, leverage the structure of networks, and acknowledge non-equilibrium. In addition, the ability and willingness to run experiments is essential to unleash a revolution. It's the only way of finding your way around a complex system. The Chinese government in its drive for rapid societal and economic change has made experimentation part of its practice. Green finance zones, different approaches to corruption, and rural land reform are all fair game for trials.[15] Experimentation is a part of the revolution approach, but it is not the entire approach. Government actions aim to be catalytic, where they trigger an avalanche of actions across a network—rather than aiming to direct each agent. Policy aims for eco-structure redesign.

Beyond Two Basins

The world is not short of wicked problems that defy easy solutions. Some are so urgent that they need answers that scale up fast; for others there is more time. Climate change necessarily implies a set of revolutionary paths, as an incremental one is no longer plausibly adequate. By all means do the incremental measures; it will do some good, and also it will prepare the ground for a more revolutionary path. And of course, at some level, all of the wicked problems are interconnected and influence each other. The loss of biodiversity is influenced by climate change. This is also the case for antibiotic resistance and water issues. Although much progress has been made, poverty and violence against women remain intractable.[16] Rising inequality is connected to most other issues in complex ways.

The image of a pebble needing to find its way from one basin to the other is convenient and illustrative. Socioeconomic and natural systems are constituted of many competing basins, shape-shifting basins that influence each other. Prigogine and Stengers, observing that societies and the natural environment are immensely complex systems, stated: "We know that such systems are highly sensitive to fluctuations. This leads to both hope and threat: hope, since even small fluctuations may grow and change the overall structure. As a result individual activity is not doomed to insignificance. On the other hand, this is also a threat, since in our universe the security of stale, permanent rules seems gone forever."[17]

The equilibrium mind-set pervading much policy making limits us to incremental change, at best. Unleashing a revolution through policy means creating those conditions where the kind of small actions Prigogine refers to might shift the overall structure. This requires policy makers to embrace complexity, so that they can make a purposeful choice between an evolution and a revolution pathway.

The Paris Climate Revolution

Paris has had its fair share of revolutions, but we refer not to 1789, 1848, or even 1968, but to 2015, when the COP21 conference was held in the City of Light. What was different?

The foundations for global climate policy were laid at the rather blandly named Conference of Parties III (COP3) in Kyoto in 1997. The basic principle was one of top-down management. Scientists would determine the global carbon budget, which was then allocated to various countries as emission limits. Much haggling occurred over the allocation, with developing countries arguing for taking into account the cumulative emissions since the industrial revolution, and the developed countries arguing for current emissions as the measure. A compromise was found and emission limits agreed upon. Then very little happened, and emissions continued to grow ineluctably.

This basic organizing principle remained the same for the next seventeen COP meetings, with little progress to show. Then COP21 happened. It was widely perceived as a breakthrough. Less widely reported, it also had a very different idea of governance that is best rendered intelligible through the perspective of Elinor Ostrom. In a 2009 report for the World Bank—coincidentally the year of her Nobel Memorial Prize—Ostrom dissects the causes of the failure of the Kyoto process: "In addition to the problem of waiting too long, 'global solutions' negotiated at a global level, if not backed up by a variety of efforts at national, regional, and local levels, however, are not guaranteed to work well." Building on her decades of patient observation of how people in a variety of systems solved their collective action problems without succumbing to the tragedy of the commons, she argues for a different model of governance.[18] She proposes a polycentric approach: the climate problem is complex and occurs at multiple levels around the world—so the solution must fit the nature of the problem. "Given the complexity and changing nature of the problems involved in coping with climate change, there are no "optimal" solutions that can be used to make substantial reductions in the level of greenhouse gases emitted into the atmosphere. A major reduction in emissions is, however, needed. The advantage of a polycentric approach is that it encourages experimental efforts at multiple levels, as well as the development of methods for assessing the benefits and costs of particular strategies adopted in one type of ecosystem and comparing these with results obtained in other ecosystems."[19]

If that sounds messy, it is because the problem is messy and beyond simple solutions. But polycentric governance is well described by Ostrom and others. Our example of the *subak* system illustrates how it works in a basic system. The Kyoto process was a simple solution to a messy problem that was unlikely to work from first principles.

The good news is that Ostrom's idea was understood by several people at the heart of the UN's climate change effort. For the 21st COP in Paris, a revolution was planned—and delivered. Instead of debating the implementation of imposed targets, in a radical reversal, countries were asked to provide their best plans for dealing with the climate crisis. The collective problem was well understood, as in any community each was asked what they could contribute. Predictably, when you added it all up, it fell short of what was required.

That is where ratcheting comes in. A ratchet is an old idea. The Romans used them in their siege engines that hurled rocks at enemy positions. Most familiar is their use in mechanical clocks, where a cogwheel with asymmetrical teeth engages with a pawl. This pawl ensures that the wheel can turn in one direction only.

In a brilliant addition, the organizers added a ratcheting mechanism to the agreement. The negotiators acknowledged that their collective effort was falling short and agreed that they would ratchet up their commitments every five years. The combination of the bottom-up submissions with the ratcheting mechanism was a revolution, from a top-down to a bottom-up process. Of course, the global level is only one dimension of the effort, and one hopes that the change in governance carries through at the national, regional, local, and family levels.

PART II

REVOLUTION POLICIES

Before we turn our attention to some examples of complexity frame policy for the climate crisis, a warning to the reader: we all have a very marked innate predilection for top-down solutions. They are simple to understand, and one pictures clearly how they might work. However, as stated in Chapter 1, they are great, but just not happening. So by all means let's continue to pursue them, but just not make them the only game in town.

Fixing the climate is actually not difficult—at least in principle. There are stacks of reports that tell you what to do. And also that it doesn't cost very much, at least much less than not doing those things. Cut coal, curb carnivores, and insulate buildings is a pretty good starter list. Fixing the emissions from the "harder-to-abate" industries, such as steel, cement, and heavy transport, is a little harder, but even there the solutions are quite well understood.[1] Even fixing emissions from other systems is not that hard. A 2018 report by an industry-sponsored group concludes that "it is technically possible to decarbonize all the harder-to-abate sectors by mid-century at a total cost of well less than 0.5% of global GDP."[2] While there is a certain satisfaction in realizing it is possible, and perhaps sharing such insights over beers, that doesn't make it happen. The track record of acting on these sorts of recommendations is poor.

The reason is that there are simply too many path dependencies that lock the system in place. Sometimes we can get a lucky break, such as when the United States curbed mercury and developed a shale gas industry concurrently. But luck is not a strategy.

These few paragraphs are indeed a warning to the reader. In the next chapters he or she will experience surges of nostalgia for the simplicity of these top-down recipes over the messiness of polycentric network-driven solutions. Experienced policy makers will agree with Elinor Ostrom that the belief in the existence of panaceas for wicked problems is dangerously distracting from real solutions. At the same time, populists the world over will continue to feed the belief in simple solutions by dangling alluring promises of easy fixes.

We must learn to loosen our penchant for panaceas and learn about revolutions through policy. Because complex systems are driven by interactions with other systems, and experience emergent behavior from within, it is clear that any solution is temporary and context-dependent. Therefore we cannot present plausible sweeping solutions, but instead examples that illustrate how to frame things differently. Framing is everything. The aim is to provide a different frame—a complementary frame—that allows for the design and the imagination of new climate solutions.

It will take some time for people to grow accustomed to and comfortable with the new frames—time we don't have. But it doesn't need to take long, as people have increasingly been primed with the urgency of climate action for a long time, and complexity resonates with our intuition, if only we let it.

6

Kicking the Coal Habit

It's complicated to project yourself into the future, when
everything that has defined you has disappeared.

JEAN-FRANÇOIS CARON

The smoking gun is coal. To fix the climate, quit coal. It is by far the
most carbon-intensive fuel, and there are cost-effective alternatives.
Yet, paradoxically, we are expanding coal capacity. Why would we do
that?

Coal enabled the industrial revolution and laid the foundations for
modern industrial development. It is a resource that is widely distributed
and could historically be mined in close proximity to the factories. That is
still largely true today. Industrial development in Vietnam, India, Poland,
and South Africa is powered with local coal. While China increasingly
brings in coal from Australia, it also has enormous domestic reserves. Almost 12,000 coal plants still power the world, with another 1,200 under
construction.[1] Deep path dependencies guarantee the resilience of coal: large
industries profit from it, bankers are used to financing the plants and comfortable with their risk profiles, power grids are designed as star networks
around them. And in some places such large-scale local projects become
an irresistible temptation as well as a convenient vehicle for graft.[2] The
social and environmental consequences were always apparent, as Charles
Dickens notes in *Bleak House* in 1852: "Smoke lowering down from
chimney-pots, making a soft black drizzle, with flakes of soot in it as big as
full-grown snow-flakes—gone into mourning, one might imagine, for the
death of the sun." But it was a price deemed worth paying for economic
growth.

There is much debate on how to address the climate crisis, but one thing
is abundantly clear. Burning coal should be firmly curbed, quickly and universally. In 2018, according to the International Energy Agency (IEA), "Coal-
fired electricity generation accounted for 30% of global CO_2 emissions. The

majority of that generation is found today in Asia, where average plants are only 12 years old, decades younger than their average economic lifetime of around 40 years." And the absolute level of emissions is still growing.[3] These arguments are widely available, and we won't further summarize them here, if only to underscore our argument that analytic considerations are largely ineffective to curb coal. In many countries the cost of renewables is rapidly coming down—and it is already economically competitive with new coal capacity in some places. Economics, however, is only one element of the story. And while the choice of power source is often portrayed as mostly a money issue, in practice many other factors weigh in. Understanding and engaging with those factors is essential to overcome the lock-in of coal. This may help pinpoint those measures that can tip the coal ecosystem into the dustbin of history. But those measures are likely to be location- and time-dependent. As is typical for complex systems, they may not be universal recipes.

Horses for Courses

In the United States, as we have seen, it was the combination of very strict mercury regulations and the emergence of a shale gas industry that nudged coal into a sharp decline. In China the construction of new coal plants has stopped, heralding a reversal of a seemingly inexorable expansion that had fueled the country's rise. Many factors contributed, but a critical trigger to break coal's growth was concern over air quality in cities. Fine particle pollution grew to record levels, and it became increasingly unsustainable to disregard public opinion and concern. Emissions from coal plants located in proximity to cities were a big factor and increased pressure to curb them. This pressure in turn allowed the government to stare down the entrenched interests of the coal industry and call a halt to its growth, setting the stage for a future decline. But while that meant stopping the construction of coal plants in China, the industrial cluster of companies that designed and built these plants and their components was not scrapped. So naturally they looked for new markets. The ensuing boom in coal plant construction in Asia, using Chinese technology and financing, is a direct consequence of these companies needing to find other outlets. Vietnam and India are expanding coal capacity at breakneck speed. And while that is framed as an economic choice and the lowest-cost option for development, it is plausibly also a product of inertia and a result of China's domestic change of heart. Interconnections are everywhere.

The Netherlands has taken yet another route to capping coal. Many of its coal plants are recent constructions. Permitted in the early years of the twenty-first century, the new 3,500 megawatts of capacity was essentially intended as diversification of the energy system. The country was overly reliant on a single source of power—its abundant domestic natural gas—and the government had been worried about a lack of alternatives. Renewables were not yet foreseen to be a viable alternative for many decades, so the government embarked on an aggressive program of new coal builds. Many years in the making, the brand-new plants finally came on stream just when coal was becoming a four-letter word. Then came the lawsuit. Urgenda, an NGO representing hundreds of citizens, sued the government for negligence in its duty of care for citizens. In 2015, to great consternation, a district court recognized its argument that the government could not advocate a minimum carbon reduction of 25–40 percent for the developed countries in international negotiations while falling well short of this level in its domestic policy. And the Netherlands, being under sea level, is particularly exposed to the danger of rising waters. An appeal by the state was resoundingly rejected in 2018. In practice the verdict implies a rapid acceleration of the closure of its shiny new coal fleet.

Every country has its story, and we could go on. What is clear is that in many cases the thrust for capping coal does not come from a straightforward top-down policy to throttle the carbon belching because it's the right thing to do. Coal's demise mostly comes from an intervention in an adjacent system, with a knock-on effect into the energy system.

But there are ways to facilitate the path when the time comes.

Mind the Miners

Jean-François Caron is the mayor of Loos-en-Gobelle, a small town in the old coal basin in the Pas de Calais of northern France. When he became the mayor in 2001, the state of the town was dismal. The last coal mine had closed in 1986, and sharing the fate of other former coal regions like Appalachia, misery set in. The soil was heavily polluted, unemployment had skyrocketed, and the men were dying of lung diseases. "The question was what to do with the town, when it had been entirely defined by coal and that no longer existed. What is our future?" asked the mayor.

Recognizing that the challenge of moving on was rooted in the identity of the region, he took an unusual step to celebrate the past. And by 2012, the entire region was declared a UNESCO World Heritage Site:

Remarkable as a landscape shaped over three centuries of coal extraction from the 1700s to the 1900s, the site consists of 109 separate components over 120,000 ha [hectares]. It features mining pits (the oldest of which dates from 1850) and lift infrastructure, slag heaps (some of which cover 90 ha and exceed 140 m in height), coal transport infrastructure, railway stations, workers' estates and mining villages including social habitat, schools, religious buildings, health and community facilities, company premises, owners and managers' houses, town halls and more. The site bears testimony to the quest to create model workers' cities from the mid 19th century to the 1960s and further illustrates a significant period in the history of industrial Europe. It documents the living conditions of workers and the solidarity to which it gave rise.[4]

Mayor Caron's conviction is that you can't invent a new future without being able to celebrate the past. Demonizing coal is counterproductive, as it negates people's past and the value of the sacrifices of their parents and grandparents.

As a green politician, Caron doesn't need to be convinced how damaging coal is to the environment, but his conviction is that to go beyond coal you need to give it its due place in history. Coal enabled the industrial revolution and helped lift billions out of poverty. Only by acknowledging and celebrating this can you move forward. Crucially, this sidesteps the political backlash and populist support that is stalling coal reform from Poland to Appalachia. Hillary Clinton is often quoted as having said, "We're going to put a lot of coal miners and coal companies out of business." Less often mentioned is the next paragraph of the speech: "And we're going to make it clear that we don't want to forget those people. Those people labored in those mines for generations, losing their health, often losing their lives to turn on our lights and power our factories. Now we've got to move away from coal and all the other fossil fuels, but I don't want to move away from the people who did the best they could to produce energy that we relied on."[5]

As in the United States, French coal consumption has been gradually dropping.[6] But both Clinton and Caron are right that without real attention to the social legacy of coal, it becomes much harder to reverse the resistance to its dominance and its stranglehold on the public psyche. Although Germany has fewer than 25,000 coal workers left, they are vocal and underpin a political shift to the far right. It remains the Achilles' heel of Germany's Energiewende. The coal miner projects an image of virility, of nostalgia for a time when a man's physical labor was a source of pride and a robot was something you read about in Asimov's fiction. Neglect the cultural and gender identity dimensions, and coal will lead to further social and environmental damage.

Follow the Money

Funding for the 1,000-plus coal plants on the drawing board overwhelmingly comes from a few countries. For instance, China's Belt and Road Initiative (BRI), a veritable twenty-first-century version of the ancient Silk Road, will be largely coal-powered. The current BRI countries account for 18 percent of global GDP and 26 percent of global carbon dioxide emissions. The level of investment that could be catalyzed through BRI is such that in the worst-case scenario, the BRI countries could account for over half of global CO_2 emissions by 2050.[7] However, the bulk of coal generating stations on the drawing board for Belt and Road countries will be anchor-funded by Chinese, Japanese, or Korean financial institutions. And while a handful of overwhelmingly European banks have adopted policies against funding either coal plants or coalmines, this is not the case in Asia.[8] Changing policy at a small number of financial institutions would redefine the coal landscape for the world. And while these banks are not independent of the political economic interests of their countries, particularly in China, their policies nevertheless represent a potential catalytic force for the world's climate. New policies in those few companies would be the end of the road for coal in the BRI.

There's another Achilles' heel for finance: impairment. Every year the management of every firm is called upon by their accountants to state the value of their assets. The accountants ask questions, audit, and then certify the logic with which the assets were valued. But what is the remaining value of a coal plant? Basically, it is determined by how much money it will make over its remaining useful life. But how long is its useful life? Under any realistic climate scenario, coal plants will either be closed in a more or less congenial way today, or more brutally in the future as the urgency further grows. Either way there is huge range of uncertainty over the value of these plants. Depreciate too fast and you decimate short-term profit and shareholder value. Depreciate too slowly and you are liable for cooking the books. In Europe $75 billion worth of coal plants have been depreciated early;[9] this is called impairment.

Just as in Urgenda in the Netherlands, expect civil society organizations to bring lawsuits against utilities and their accountants for misrepresenting the value of the coal plants on their books—and thus artificially pumping up the value of their shares. This may play out in different ways. Shareholders will see some value evaporate, and some of these risks are implicitly guaranteed by governments. This is because many banks understand the impairment risk and ask for government guarantees before agreeing to finance a coal plant. So the taxpayer will get stuck with some of the bill for impairment. On the flip side, if a coal plant is fully depreciated, it rather undermines the argument of a utility to require compensation for closing it early. Catch-22.

Either way, like mercury regulations, impairment has the potential to destabilize the coal industry with some fairly limited interventions. Following the Urgenda lead, a couple of well-aimed and successful lawsuits against utility management would resonate around the world.

No Panacea

The UK was the first nation to commit to ending coal use—by 2025—and the electricity generated by coal there has already fallen from 40 percent to 2 percent since 2012. The Powering Past Coal alliance announced at COP23 assembled countries willing to make commitments.[10] Germany has announced a schedule to close all its coal plants by 2038.[11] And yet emissions from coal keep rising inexorably. Curbing coal remains one of the most obvious and yet one of the hardest elements of dealing with the climate crisis. But facing down this challenge starts by recognizing that coal is deeply interconnected with other systems. Only by identifying those interconnections and dealing with them at the appropriate scale can progress be made. In many cases the efforts involved are relatively limited: either through a catalytic intervention such as the U.S. mercury regulation, social programs as in the Pas de Calais, or new investment policies at a few financial institutions. The measures will be different in each case, and most certainly will need to evolve over time. Moving beyond a standard policy framing to take into account the complex system nature of the problem can open the way to solutions at various scales.

⑦

Wrong for the Right Reason?

It's saying that there are two ways of solving problems, and just for the sake of discussion, let's call one "incremental" and one "revolutionary," and they're both great. The incremental approach is what we typically do, we say, "What is the problem?" Let's specify all the components of the problem. Let's find the tools that are necessary to solve the problem and so on, and that's great, but there's another way of doing it, and that is setting yourself a vastly harder problem.

DAVID KRAKAUER, PRESIDENT OF THE SANTA FE INSTITUTE

Challenging times make for strange bedfellows. A dour German conservative politician has flipped the views of many environmentalists; their opinions on nuclear power have gone from revulsion to the kind of regret one might have for a former lover too mindlessly spurned. And it took only four months.

The Fukushima nuclear disaster started in March 2011, and within days the German cabinet shut down the seven oldest atomic power plants and ordered detailed audits of the entire fleet. On June 6 the cabinet decided to exit nuclear power altogether and jettison all plants by 2022. The decision was ratified quickly and overwhelmingly by both chambers of parliament and became law by August 6. In a mere twelve weeks Germany had made a full U-turn on decades of energy policy. Eight nuclear plants lost their licenses immediately, and the remaining four would be shut down over the following decade as part of the "managed" rush toward a full nuclear exit.

The Energiewende

All this happened while Germany was—and still is—in the midst of its Energiewende, a grand experiment to cut its energy consumption in half by midcentury and cut greenhouse gas emissions of the remaining production to near zero.[1] "Energiewende" has caught on as a way to describe a fundamental restructuring of the entire energy system and energy policy. It often goes hand in hand with a whim of overtly political ambition, if not outright hubris. First set a target—lay out an overall vision—and then strive for all the pieces to fall into place. There is indeed a fine line between ambition and hubris, and there is plenty of room for spinmeisters on both sides. "Shooting for the stars" is fine, as the saying goes, but it is clear that there are plenty who will declare "reaching the moon" an abject failure.

Opinions are split on the meaning of the term "Energiewende." Purists will insist that, like many compound German terms, it eludes translation. The term is suitably vague to accommodate multiple interpretations, but its literal meaning conveys raw ambition. Translated literally, it means "energy turnaround" or "energy U-turn." The catchall term, thus, covers a vast array of diverse approaches. They range from concrete policy interventions with measurable outcomes, like the adoption of ambitious feed-in tariffs that have guaranteed users a return on their investment in solar power, to a broad grab bag of interventions, like the creation of energy-focused institutions and think tanks. Those constitute a veritable ecosystem for debate, challenge, and innovation. That's good as far as it goes, but, of course, the outcomes are difficult to measure, and more difficult to defend, when challenged with a sole focus on benefits and costs.

Whatever it means, the Energiewende has, in fact, helped transform Germany's energy sector at a substantial scale. It has also come at a steep cost to incumbents and consumers. Along with China's investment in a vast production capacity, this feed-in tariff policy is widely credited with driving down the cost of photovoltaic panels much faster than any analyst had imagined possible.[2] German households now pay one-third more for electricity than they did before the feed-in tariffs.[3] Industry, meanwhile, emerged largely unscathed. Its power prices were ring-fenced and insulated from most increases. While this helped create a degree of indifference on the part of business, public support, remarkably, has remained consistently high across political affiliations. Surveys show that the majority of respondents consistently take the position that the Energiewende should be further accelerated.[4] It has helped to have Chancellor Angela Merkel, a conservative politician and former environment minister, at the helm, deftly navigating the politics and dodging the reefs.

Not Uncontested

While this is the majority opinion, there is obviously no shortage of pundits who are critical of the Energiewende. Some point to the large subsidies as wasteful,[5] others to the unfairness of Germany needing to bear all the cost on its own as the first mover while other countries dithered; other voices dispute the need to move as quickly on climate policy. Still others point to the disappointing outcome, as Germany's emissions have hardly budged. These arguments are not much different from those in other countries. What is unique is the consistent strong majority support for positive action on the Energiewende. And seen through a revolution lens, we'll see that the policy itself helped build further support, creating a positive feedback loop. Our concern here is not so much with whether German policy is optimal or even a good idea, but to use it as a real-life illustration of dealing with path dependencies in a purposeful way through policy intervention.

Nuclear power represented almost a quarter of Germany's power production—and emitted zero greenhouse gases. Shutting down the plants made the task of eliminating emissions in Germany suddenly a whole lot harder. Yet the government decided that the nuclear exit would be absorbed and not influence the intended targets. In 2011 emissions had already been cut by a quarter, although greatly assisted by the shutdown of the toxic fume–belching East German power system, with its fleet of aging coal plants. On the demand side, consumption had barely budged.[6] Courageously, Germany would take the loss of the emission-free nukes on the chin and retain its original reduction targets.

Real Men Don't Do Renewables

In the early days of the Energiewende, in the boardrooms of the large German utilities, the slow rise of solar and wind power was observed with skepticism. RWE, E.ON and their brethren had reemerged from the ruins of World War II, growing steadily to underpin the German economic miracle. From their headquarters in the industrial heartland in the Ruhr, these companies had been part of the fabric of German industry for the better part of a century. Closely associated with local mines, large centralized coal plants formed the bedrock of their operation, complemented with a fleet of hulking nuclear power plants. A smaller number of gas plants completed the system. Power was provided centrally, with huge capital investments managed by a small group of well-connected directors—virtually all male until this day. In this robust engineering culture, problems were resolved either by technical solutions or by financial maneuvers with a small network of bankers, as well as government officials with whom the directors had close ties. The system worked extraordinarily well, providing reliable if not rock bottom–priced power,[7] underpinned by a hefty dose of subsidies for both coal and nuclear.[8] When the Energiewende started in the early

1990, the men (*sic*) at the power companies considered it a distant and curious development, very much at the periphery of their world.

There was much to dislike: unfamiliar technologies to integrate into their networks, new competitors on the horizon, and competition for their comfortable subsidy streams. But most importantly, it was not their world. The technical challenge of renewables simply did not interest the engineers. Their traditional plants were largely custom-designed units, involving thousands upon thousands of hours of professional engineering, each with new challenges that could be creatively resolved. The renewable kits, however, came shrink-wrapped out of a distant factory. The only thing to be done was to install them, and do so with great cost efficiency. For companies whose core task for decades had been to manage multiyear large-scale engineering projects, this was both unfamiliar and deeply unattractive. No fun.

As we'll elaborate in Chapter 10, it is easy to underestimate the role of such soft factors in corporate choices. While economists and MBAs describe corporate policy in terms of financial metrics, reality is much more multifaceted. Companies acquire competences, habits, and even a language optimized around what they do. And real men loved large centralized power plants. And yes—it's mostly a gender thing. That is what they liked to do and were comfortable with as a business. They locked into a set of technologies, habits, and the relationships that fit with big plants.

Then in 2011 came the sudden nuclear withdrawal. Executed with lightning speed, along with the surprisingly competitive cost of wind and solar power, at the stroke of a pen it became impossible for the lumbering energy giants to ignore the Energiewende. To add insult to injury, the market capitalizations of German utilities continued to erode rapidly. By 2013 incumbent E.ON had shed three quarters of its stock market value.[9]

Parallel Narratives

Coming so quickly after the nuclear disaster at Fukushima in Japan, the nuclear exit was widely reported (and much criticized) as Merkel shooting from the hip. In the immediate aftermath of the disaster, the chancellor had contacted a number of CEOs of the major utilities to ask whether each of them might be prepared to shut down one of their reactors, at least temporarily. The responses were evasive. Greeted with noisy antinuclear protesters at a press conference in the Baden Württemberg, Angela Merkel stunned her audience by stating that she now realized that the risk of nuclear power had been structurally underestimated. The protesters fell silent. It was as if "the pope were suddenly advocating the use of birth control pills."[10]

But here is another version of the same story. This frames the decision to exit nuclear as a catalyst for a systems change, for a revolution, rather than simply to terminate one technology option.

Germany's opposition to nuclear power has deep and strong roots. Since the late 1970s growing protests have been voiced against expanding the nuclear industry, under the red and yellow icon of "Atom Energie? Nein danke" (Nuclear energy? No thanks). In 1973 a start was made to build a new generation of plutonium-powered plants, at Kalkar. Eighteen years later, it was turned into an amusement park.[11] It is a Disneyland-lite where the vast cooling tower for the nuclear reactors now functioned as "Echoland." However, there was nothing playful about the over four billion euros spent before the decision was made to quit. Politicians in Germany knew there was strong explicit and latent opposition to nuclear power. When the decision was taken to exit nuclear, one of the officials at the ministry remarked, "There are historic moments that you have to seize if you want to achieve something."[12] This was such a moment.

It was a moment to stop the nuclear industry, certainly. But our perspective in this book is about addressing path dependencies to create system change. From this point of view, this was also a moment to slash a critical path dependency that stood in the way of further unshackling Germany's energy transition. The move took away an essential leg of the centralized power system, rendering it unstable. And complexity science shows us that only systems out of equilibrium can change in meaningful ways. It signaled to the incumbent power companies that their business models based on centralized provision of power, led by their armies of engineers, were no longer viable, nor welcome. As one of the initiators of the Energiewende, Chancellor Merkel had seized the opportunity to slash one of the critical ties of the energy system to its own past.

Sure, there was a big cost, both economically and environmentally. With nuclear power plants—as with renewables—almost all of the costs are spent up front. In contrast with coal or gas plants, which require a constant and costly supply of fuel, once you have sunk the cost of construction, nuclear plants are quite inexpensive to run on a day-to-day basis. Faced with a high bill for decarbonizing the power system, having nuclear power with almost zero marginal cost is economically very useful. On top of that, the early decommissioning would also trigger the huge expenditure to dismantle the radioactive buildings. Although that cost was coming someday, having it earlier rather than later is onerous.[13]

From a standard economics perspective, the early closure was not an efficient use of capital resources.

The environmental cost stemmed from the fact that nuclear power does not produce greenhouse gases along with the electricity. As such, nuclear power was an important contribution to the Energiewende, as it would provide some breathing room to develop a renewable infrastructure, while the nukes continued to deliver their zero-carbon electrons. While many German NGOs had been firmly opposed to building new nuclear capacity, others voiced considerable concerns about closing existing plants early. And they would be proven right. The rapid decrease in German carbon emissions was to grind to a halt, as mainly coal capacity took over some of the role of the nuclear fleet.

Some will conclude that these liberal Germans, with their antinuclear stickers, have achieved less than the Americans in terms of emissions reductions. The facts are more ambiguous: U.S. emissions have actually risen by 8 percent since the reference year 1990, while German emissions have fallen by 24 percent. And the average American emits a lot more carbon than the average German.[14] On the other hand, German emissions have not fallen much since kicking off the Energiewende in 2010, while American emissions have also been essentially flat. The problem with such statistics is that they implicitly assume that if the system changes, falling emissions will immediately demonstrate this. In this chapter we have argue that a long period of adjustment has been necessary to break path dependencies and change the structure of the system. Those changes are not immediately evidenced by proportionally falling emissions. That is certainly a reason to be skeptical, but our understanding of complex systems also means that we need to look beyond the linear relation between change and measurable output.

There are two signs of a deep shift in the German system. After only a few months, a commission proposed a plan to close all coal plants on a tight schedule. This would have been politically unthinkable only a few years ago. Also, the German government has signaled its support for a carbon-neutral Europe by 2050, another significant shift in position.[15]

There you have it: economically suboptimal and environmentally dubious. However, seen through our lens of revolution policy, it may well make all the difference in the end. While it is uncertain, the value can be very large—as it could prove to have been the critical ingredient that enabled systemic change. In this case higher cost and more emissions are traded off against an acceleration of the Energiewende. But we cannot be sure because we don't (yet) have the tools for computing the value of cutting a path dependency. This is an equation economists cannot solve, but it is one that policy makers need to tackle to achieve a revolution. It should surprise no one that this argumentation was not used to defend the measure publicly. Building on the decades-long suspicion of the safety of nuclear plants, this became the chosen argument. While the argument was accurate, it highlights an ethical problem with the kind of system change policies we advocate. If the argument is not made, or cannot be made directly, are policy makers being manipulative? We interpret the effect of the nuclear exit primarily as a way of unseating the lock-in of traditional power companies. But if that had indeed been the intent, should it have been presented as such?

⑧

Norms Policy

Those are my principles, and if you don't like them . . . well,
I have others.

GROUCHO MARX

If you know someone who knows someone who knows someone who is
overweight, you are more likely to gain weight. A precise look at weight
gain and loss across networks of friends and colleagues shows that being
overweight spreads very much like a virus.[1] It's not a virus, of course.
Imagine you sit around a dinner table in a restaurant, deliberating whether
to skip dessert. It only takes someone to say "Come on, don't be a spoiler
and join us" to sway your opinion and indulge. Social norms are contagious, much like diseases. And just like diseases, there are different degrees
of contagion, depending on which norms are involved and the structure of
the network across which they spread.

It turns out that values, social norms, and principles depend a lot on the
context. As in biology, they mate and evolve depending on their connections and their environment.[2] Better understanding how norms that are fa-
vorable to the environment are formed and how they spread is a big opportunity for more effective climate policies.

Violence Interrupters

"So there are these two marching bands got this beef going on," recounts
one with delight, through an open window of Mr. Barksdale's car, "and
they got knives and pit-bulls." Mr. Barksdale is a "violence interrupter" in
Baltimore talking to one of his thirty colleagues at Safe Streets. He knows
what he is talking about, as he spent ten years in prison for selling drugs.[3]
Although Baltimore's violence record is dismal, Safe Streets is following a
model that has been successful in other cities.

The idea is to interrupt the network propagation of social norms. Violence is contagious.[4] Violent people are bad people, so you should lock them up, right? What about if you've caught the bug from someone else? Whose fault is that? Should we lock up the person who made you violent? But what if it is a whole network structure that enabled the contagion? How can you punish a network? Gary Slutkin is an epidemiologist by training who worked across multiple continents to deal with various diseases. Then he took an interest in violence in his home of Chicago and was struck by the similarities: "This violence had fulfilled the criteria of the population characteristics of a contagious disease."[5] We are not against punishing violent people, but the bigger policy goal is surely to have less violence in society. In this case locking up violent criminals may be only slightly more effective than locking up people with the flu.[6] With the assistance of the University of Illinois in Chicago, Dr. Slutkin founded his own NGO, Cure Violence. It advocates tackling violence like a disease. This means identifying the sources and degree of contagion and understanding the characteristics of the networks across which it spreads. Only then can you intervene by changing the structure of the networks, or isolating particularly strong hubs of contagion. One particular strategy is to recruit "violence interrupters." These are individuals who have strong relationships in the community and often a violent history themselves. Through their position in the network and their inclination, they can change the contagion effect. The approach is promising and has been reproduced in other cities—in Baltimore, for example.

Kicking the Habit

In the 1990s Inga Dóra Sigfúsdóttir, a young research assistant at the University of Iceland in Reykjavik, surveyed the drug use of fourteen- to sixteen-year-olds. The results were alarming. Twenty-five percent were smoking every day, and 40 percent had gotten drunk in the last month. Crucially, the researchers went beyond the average statistics and looked at factors that might explain differences. Practicing sports and being out late at night, as well as the quality of the relationship with parents, all had an impact. In 2017 The Atlantic reported: "Today, Iceland tops the European table for the cleanest-living teens. The percentage of 15- and 16-year-olds who had been drunk in the previous month plummeted from 42 percent in 1998 to 5 percent in 2016. The percentage who have ever used cannabis is down from 17 percent to 7 percent. Those smoking cigarettes every day fell from 23 percent to just 3 percent."[7]

Previously there had been all sorts of education programs to point out the dangers of various substances, but they weren't working. Just explaining things did not lead to change. Iceland shows how people's physical addictions can be changed, by changing the context surrounding them. Curfews, sports, life-skills training, art, financial assistance, spending time with parents all contributed to fundamentally changing the network surrounding each child, redesigning the influences that drove their behavior.

Infectious Solar

Much has been written about the economic merits of the feed-in tariffs for solar power. Introduced in 1991, they bought back power from those who had installed solar panels at a level commensurate with the cost of the panels. As the prices slid down the cost curve, the feed-in tariffs were progressively adjusted downward. This is a remarkably sophisticated policy, as it required the rules and regulations to be constantly adjusted to a moving target. Not only prices influence our behavior. They do, but there is another story visible through a revolution lens, a story of contagion.[8]

When you put solar panels on your roof, your neighbors will interrogate you and lob comments at you. They might say they object or find them ugly. Or they may ask how long it takes to earn the cost of the panels back through electricity savings. And they may well ask about you, for example, whether you are now also a vegan, and ask about that Porsche in your garage—questions, in short, about your identity.

People decide to go solar by talking to their neighbors and friends. The result of all the interactions and referrals is a consistent pattern of solar adoption. Companies like SolarCity in the United States have reacted to this by appointing "solar ambassadors" who function as highly connected nodes in the network of solar adopters. The models that are used to understand this phenomenon have the same structure as those used by epidemiologists to map diseases. It essentially consists of two parameters: how many people interact with each other tightly enough for contagion, and also how contagious a particular disease might be. Mitigating an epidemic consists in tweaking those two parameters: the structure of the network and the infection rate. The same approach can be taken for spreading adoption of solar panels.

The U.S. city that takes the top spot for solar contagiousness is Fort Collins, Colorado, where 69 percent of its solar customers were referred by a friend, closely followed by Kona, Hawaii (64 percent), and Gloucester Township, New Jersey (62 percent). Colorado's prominence on this list of contagious solar cities may not be a coincidence, but rather stems from the state's tight-knit social dynamics. Colorado has one of the highest rates of people who "talk to their neighbors" (88.2 percent) and are "active in their neighborhood" (10.8 percent).[9] Fort Collins in particular has one of the highest rates of volunteerism (38.2 percent) in the country among midsize cities[10]—implying the kind of community cohesion and social capital that are conducive to encouraging each other to go solar. The contagion is clearly local and doesn't jump from city to city, and sometimes not from neighborhood to neighborhood. Similar studies have been done in other countries. In the German city of Wiesbaden a study showed—unsurprisingly—that contagion is greater in higher-income neighborhoods, and by how much.[11] It requires dedicated action to create new infection centers from which adoption can spread.

The revolution story on solar PV complements the standard narratives on prices driving solar adoption. While price certainly has an effect, it remains shrouded in statistics that merely look at averages.[12] Average statistics erase any information about the structure of the network. As we explored in Chapter 3, the structure of the network is the essential ingredient for understanding its behavior. Fortunately, geospatial statistics on solar panel adoption now make it possible to see the contagion effect. The result is that the German feed-in tariffs were much more than an economic mechanism; they were also a norms policy.[13] Because of the uniquely personal nature of solar power, and its high visibility on rooftops, this policy influenced opinions and norms about the Energiewende, while at the same time ensuring its progress. This dual nature of a policy that promulgates new norms about renewable energy, while at the same time reducing costs and increasing penetration, makes this a systemic change policy—a revolution policy.

While we understand the connection between social norms and the spread of solar panels, it is beyond our current capability to demonstrate quantitatively how much this has contributed to the impressive sustained support for the Energiewende. Doubtless there are other contributing factors, such as German culture, confidence from its economic success, and other unknown factors. But recognizing that purposefully designing policies that explicitly aim to change the underlying network structure is an important tool for those policy makers who want to realize systemic change.

Beans for Beef

"With one dietary change, the U.S. could almost meet greenhouse-gas emission goals."[14] If everyone in the United States replaced beef with beans in their diet, it would achieve about half of the reductions needed to meet its 2020 greenhouse gas target. Cows emit lots of methane, a heavy greenhouse gas with eighty-four times the impact of carbon dioxide.[15] *The Economist* recently quipped that "if cows were a country, the United Herds of Earth would be the planet's third largest greenhouse-gas emitter."[16] Between actress Pamela Anderson and ice hockey player Mike Zigomanis, A to Z lists of celebrities testify to reduced meat consumption. A great idea, right? Except that saying so doesn't make it so. It also requires a theory for how to get there.

The proportion of vegetarians and vegans varies greatly between countries.[17] Unsurprisingly, in India about a third of the population does not eat any meat. But several other countries have double-digit abstainers, a diverse bunch that includes Australia, Taiwan, Germany, Mexico, and Brazil. Remarkably, Australia also takes the bronze medal for the highest meat consumption, which, along with the high proportion of vegetarians, suggests that social segregation plays a role. The mechanism is surprisingly local, as the high carnivorous preferences of the Danes and the Dutch contrast strongly with those of the neighboring Germans and Swedes.

We don't wake up one morning and suddenly decide to stop eating meat. More likely it comes about as a result of what a trusted friend has told you, or if your son comes home from school one day resolved to quit meat and convinces the family to follow suit. Like the adoption of solar panels or the spread of obesity, it is plausible that it is a contagious process. But it has not been studied as such, yet. Reducing meat consumption is a highly desirable climate policy, and if it is indeed a network phenomenon, then getting to the bottom of the mechanism that drives the spreading becomes essential. Complexity scientists would be able to model the network and experiment with intervention that might accelerate the effect. But what they need is data from individuals, not the average data that is generally available. For solar panels the network was geographical, so that knowing when and where PV went up was enough to map out the underlying network structure. Vegetarianism is more like obesity, where you need to have a map of the relationships between individual people and their acquaintances, colleagues, family, or close friends as well as when each of them stopped eating meat. It takes some effort to collect this data, but the payoff could be revolutionary.

A vegan policy would come at little or no cost for the government or for individuals. All it takes is a little bit of studying to identify the contagion mechanism, and it might just spread like wildfire. As the sale of meat substitutes is booming, the meat industry is feeling the winds of change and fighting back. "The word meat, to me, should mean a product from a live animal," said Jim Dinklage, a rancher and the president of the Independent Cattlemen of Nebraska, who has testified in support of meat-labeling legislation in his state.[18] The meat lobbyists suspect that labeling is a key driver for meat substitution. It may well be one of the drivers, or they may have it the wrong way around, but in any case the underlying network effect remains to be discovered. And like an epidemic—but a good one—it would be harder to slow down once the contagion started.

The Calm before the Tipping Point

Systemic change is in principle directionless. Systems don't naturally improve when unshackled; they can just as well deteriorate. Which means that when you break path dependencies, you can also end up in a really bad place. An important way to mitigate this risk is through norms policies. This is tricky, as norms coevolve with the system.

Skeptics point to the disappointing emissions reductions in Germany as evidence of failure. That is true from an evolution perspective, not a revolution one—at least so far. Before a sudden change becomes possible, all sorts of parts in the system need to be rejigged. One reason there are multiple stories about the Energiewende is that big change begets turbulence. As with boiling water or freezing Corona beer, this implies that multiple states exist at the same time. And multiple states beget multiple stories. We distilled a few of the underlying trends, such as the conflicting stories about the nuclear exit, which looks nonsensical in the standard paradigm. But as a catalyst for change, knocking the established utilities out of the zone of comfort, it might just have fit the bill. This would have been difficult without decades of "Nein Danke" campaigning, and the tragic opportunity of Fukushima, but it has arguably enabled the system to cut some of its path dependencies. The German—and ensuing global—solar miracle is a testament to the power of learning by doing, in the tacit dance of a manufactured German market and China's industrial capacity. But the Energiewende also involves identity engineering—solar PV touches personal norms in a way wind power does not. German solar policy is also a norms policy, a form of ecosystem design where the new norms in turn help underpin support for the broader Energiewende.

Participation, identity, and the spread of social norms are essential ingredients for rapid systemic change. Norms policies need to be an essential part of the climate toolkit. Just not the propaganda kind, but the kind that leverages network effects purposefully.

Next we turn to the manufacturing of social norms.

9

Kicking the Consumption Habit

> Economics are the method: the object is to change the soul.
> MARGARET THATCHER

The Social Norms Factory

There is no doubt that our addiction to consumption is one of the Gordian knots at the core of the climate crisis. There is no plausible scenario in which there is enough supply of clean energy to allow everyone on the planet access to Western levels of consumption. Other constraints prohibit that, such as declining stocks of raw materials, biodiversity, and so on.[1] This is often glossed over in discussions on dealing with the climate crisis, which then merely becomes a problem of dirty supply that needs to be cleaned up. The insinuation is that if we changed every brown electron into a green one and every drop of oil into hydrogen, then modern materialistic lifestyles would become accessible for everyone. If only. The supply of energy undoubtedly needs to be cleaned up, but an even deeper transformation is required on the demand side.[2] That's not immediately appealing, as people are loath to give up what they have, and even more reluctant to reconsider their aspirations.

But our current extreme levels consumption are more of an acquired behavior—an addiction, really. Sure people are greedy. They have desired nicer furs or jewelry since the dawn of time. Research shows that this is balanced by social factors, such as the value of cooperating and sharing.[3] The intriguing question becomes how we can reduce our addiction to consumption. Once you no longer believe in Santa Claus, you don't miss him. You only feel bad about consuming less of something if you desire it. If that desire decreases, it is no longer a sacrifice. People who have decreased or eliminated their meat consumption will recognize this. What seemed like a sacrifice ahead of time turns out to harbor few regrets. Your body adjusts, so that you crave less meat—and even starts to have trouble digesting it. In complexity terms, your system has moved from one basin to another, where it is satisfied and largely walled off from returning to the previous basin. Once you have the consumer norms of the new basin, you no longer hark back to those of the old basin. The system will have changed.

Mass consumption is relatively new, and John Maynard Keynes famously reflected on what many expected the future of consumption would become, as viewed from the opening decades of the twentieth century. He predicted that "the love of money as a possession—as distinguished from the love of money as a means to the enjoyments and realities of life—will be recognized for what it is, a somewhat disgusting morbidity, one of those semi-criminal, semi-pathological propensities which one hands over with a shudder to the specialists in mental disease."[4] This reflected the conviction that above a certain—by our standards modest—income level, people would turn to the "enjoyments of life." They would spend their time with friends, make or enjoy art, and savor nature. But it turned out differently; the pursuit of money and material consumption has become a core expression of identity and enjoyment of life.

That we know how to manufacture addictions is starkly illustrated by the revival of opium addiction in the United States: "Since 1999, two hundred thousand Americans have died from overdoses related to OxyContin and other prescription opioids. Many addicts, finding prescription painkillers too expensive or too difficult to obtain, have turned to heroin. According to the American Society of Addiction Medicine, four out of five people who try heroin today started with prescription painkillers."[5] OxyContin is nothing less than a synthetic form of opium. And it is a gateway drug to heroin. The brilliant innovation enabling this new wave of consumption was the development of a pill that releases the drug over time. The company that had developed the product successfully lobbied the U.S. Food and Drug Administration to qualify it as nonaddictive. In a linguistic sleight of hand, the drug was qualified as pseudo-addictive. The solution to the onset of OxyContin addiction was simply to take more OxyContin. However, if their prescription had run out, many turned to heroin to service their "pseudo-addiction." During the drug epidemic of the 1970s, 3,000 people a year in the United States died from overdoses; thanks to powerful marketing, in 2017 there were 70,000. Since its release in 1995, the blockbuster drug has generated over $35 billion in revenue for its owners.[6]

Marketing opioids is an extreme case, since the product itself is also physically addictive. Still, it illustrates the dynamic through which companies are compelled to invent new markets all the time, to serve their own addiction to profits. This in turn is baked into the very design of the institution that is the firm. Global marketing expenditure rises inexorably, pegged at $1.3 trillion in 2018.[7] Marketing is not bad in itself; it only becomes negative when it serves needs manufactured to sustain the firm, and not a societal goal. So we know how to manufacture addictions, but do we know how to rid ourselves of them? From a complex systems perspective, people's individual decisions are influenced through the network around them, but also by their individual characteristics. We tend to overemphasize the individual preference and underestimate the power of the surrounding network.

The reader may well feel some uneasiness that the language of addiction is perhaps too strong to characterize consumption, that desiring the newest fashion is far removed from craving the next slug of cheap booze. Yet our consumption habits are deeply destructive to the natural systems that support us and at the same time do not much improve our sense of well-being. As with reaching for the next cigarette, we keep consuming although we know, can know, or should know that it is not good for us. Study after study shows that beyond a (modest) level of material wealth, further increases rapidly lose their effect in making us happier. In some countries measures of well-being have even gone down while economic indicators went up.[8]

In complexity terms, addiction is a particularly strong form of path dependency. There is addiction of the individual agent, as with opiates—but also collective addictions that come about through strong network effects. We've seen how obesity spreads through networks; this also holds for consumption. We use "addiction" for the two different effects, at the individual and at the collective level. Often we tend to overestimate the role of the individual preferences. From a climate policy perspective, we are interested in changing the network effects that lead to overconsumption.

In the previous chapter we described how network effects could lead to a revolution in the adoption of solar panels. But there is a deeper question of how the networks that shape our consumer preferences come about in the first place. They are not a natural feature of the world, but the result of design—like designing the opioid market. This is a good thing, because it means we can change them. Not only can we use network effects to engender change, but we can also tweak the creation of these networks in the first place, thereby changing the nature of the contagion.

How can we reengineer the factories that manufacture our social norms into ones that are more in line with the boundary conditions of the earth's climate? And do so without limiting individuals' freedom of choice?

The mechanisms that lead to the emergence of our social norms include institutions, regulations, and infrastructure. These influence the choices of the individual agents whose preferences coevolve with the system around them. This approach is different from making rules that force people to change their preferences, such as banning sugar or taxing carbon. Such top-down interventions have their place, but they are hard to achieve and will be opposed by those who advocate full freedom of choice. As in all complex systems, there is no universal solution, no panacea. As an illustration of a few ways we can tweak the emergence of social norms, we look at two examples of interventions that have the potential to reengineer the social norms factory. The first describes how changes in the institutional design of the corporation could change social norms, and the second example looks at the paradoxical state of drug regulations.

Tweaking the Emergence of Social Norms

"We started a very small factory in Bangladesh, producing a very small cup of yogurt with the most fortified product at the lowest cost ever. That's going to be a great thing for everyone." In a classic French accent, Danone CEO Emmanuel Faber describes the product of deep reflection by a company deemed to be at the forefront of corporate responsibility. Then he tells the story of a community meeting in a small village where a young woman asked a question from the very back of the tent. The audience had just heard expounded the benefits of the new yogurt for children's health. She said: "Hello, I'm Yamina. I wanted to ask you a question. Today, I'm doing my yoghurt myself with the milk that my neighbor is giving to me. She is giving me this milk because I take care of her children, while she is away in the fields. So I understand your yoghurt is great for my kids, but if I buy your yoghurt, what is she going to do with her kids? And so I will not buy your yoghurt." Faber observed that he had never encountered a more difficult question. It stumped the CEO, as it undercut the essence of the logic of consumption.[9] One of the changes that Danone is implementing to accommodate this kind of challenge is the new structure of "for-benefit corporations."

When a young Italian wants to start an enterprise, she is no longer forced to choose whether it is for profit or for charity. Those are the two options society offers entrepreneurs with an itch. In Italy since 2015, there is now a third option: a "società benefit."[10] This somewhat awkward linguistic concoction betrays the influence of the American "for-benefit corporation." In the United States, this legal structure now exists in over half the states, remarkably adopted with vanishingly rare bi-partisan support. In addition to a financial return, these organizations have a legal duty to serve a general public benefit, but one that they define themselves. In other countries, you are forced to choose one or the other. But the United States is solving this Manichean choice, albeit slowly.[11]

The for-benefit-corporation is a new legal entity form, first launched in Maryland in 2010. The primary goal of traditional corporations is profit maximization. While this is not a legal requirement in most jurisdictions, it is so much part of the culture that the narrative has for all practical purposes become welded to the structure. For-benefit corporations allow companies to create a different narrative, by formulating their goals in term of societal or environmental goals—creating some public benefit. It is still fundamentally a for-profit structure, with all the benefits of the commercial model, but one that defines a new asset class. This allows mission-driven investors such as some pension funds or wealthy individuals to have clearer options. It also provides market differentiation with customers. Critically, the effect has a much more revolutionary potential than adding sustainability or circularity measures to an existing for-profit corporation, as it will always remain something incremental to the core, instead of the core itself.

Corporations were not meant to be just profit-maximizing, but that is what they have become over the decades. Lynn Stout made a name for herself at Cornell University by deflating the myth of the shareholder.[12] A charismatic and passionate professor, she describes this myth as an ideology, almost a religion, albeit one that has no basis in law or practice. Contrary to popular belief, shareholders actually do not own even corporations; they simply have a contract with the company, but do not possess it. She also points out the curious fact that the advent of this ideology went hand in hand with shorter life spans and lower returns for corporations, stimulating a shortsighted focus on short-term earnings. Knowing that companies do not exist intrinsically to service the interests of the shareholder is useful, but those beliefs have become locked in quite deeply. The list of Italian for-benefit companies is dominated by health-care, food, and nature conservation entities.

Legal entity structures evolved for different reasons. The Amsterdam-based East India Company (the VOC) is generally credited with being the first such commercial entity, back in 1602. Its purpose was simply risk management. Both the risk and the margin on the spice trade with the East were very high. Pooling risk and profit worked for everyone. From this narrow beginning the idea spread globally and evolved over the centuries into the modern corporation. Its raison d'être increasingly evolved into profit maximization for its shareholders.

A subtle but important side effect of the particular structure that evolved is that the corporation not only aimed at meeting the needs of its shareholders, but also at manufacturing new consumer needs and new scarcity where there were none before. Finding or creating new markets is the name of the game. The creation of the opiate mass market in the United States is a particularly perverse example of a company manufacturing a need where there was none, with great damage. Smartphones are an example of manufacturing a need where there was none, with great benefit. Innovating products is neither intrinsically good nor bad, but that is exactly the problem. It requires judgment to decide. Judgment by the market is clearly not sufficient, as it is too easy to manufacture another consumption addiction.

Instead of letting regulators perform a triage function between good and bad addictions, for-benefit corporations have the potential to create a new asset class that embodies that function. The new identity would signal this to customers and investors with greater clarity than what a couple of decades of the corporate responsibility movement have been able to deliver. In fact, Emmanuel Faber, Danone's CEO, is deeply engaged in the process of converting his company into a for-benefit corporation. Where the legislation exists, such as in the United States, Danone is using that. Where it doesn't, it is applying for a certification scheme that provides some of the features of the full institutional change.[13] The certification is a distant second choice, as it involves an agent external to the system passing judgment, while the full-blown legal entity structure will plausibly allow the emergence of new ecosystems. Many companies may choose not to play this game—and signal their allegiance to the current model by staying in their legal structure. But it will be clearer for consumers, investors, and staff.

If one defines the benefit as a better weapon, you could register that as a for-benefit corporation. The legal entity structure does not prescribe any norms for the company to follow. It is like a new playground where the kids can design their own games. In practice, as expected, it is the companies that felt boxed in by the narratives associated with for-profit companies who made the jump. Famously, the first company to register was Patagonia, an outdoor clothing company with a mission that expresses itself better in the new structure. Founder Yves Chouinard articulated his vision in a book evocatively titled *Let My People Go Surfing*. The company encourages its customers to trade in used items and repurposes them for resale. In Italy dozens of companies have embraced the idea, ranging from service providers to health food manufacturers.[14] Dani SpA is in the environmentally challenging business of leather tanning, but is acquiring a reputation for being one of the leading green companies. Some for-benefit corporations address environmental and climate challenges head-on, but our interest here is in their potential to drive the emergence of new social norms.

Changing social norms occurs in networks, as agents influence each other to define their preferences. But the shape of those networks is strongly shaped by the choices of institutions. So institutions such as the legal frame-work for corporations matter a lot for the kind of norms that end up emerging in society. For-benefit corporations provide more paths to allow managers, investors, and customers to express their own moral preferences, for which currently there are fewer outlets. This extends to changing in-centives away from profit maximization, as this inevitably feeds spiraling consumption as new markets are invented. A for-benefit corporation might, for example, cap its profit or shrink its business when its goals are achieved. Today there is no vehicle or investment asset class in which such approaches are possible.

Our analysis here is not about any particular political agenda. It is about applying insights into the complex nature of the systems that surround us and understanding better how these lead to certain collective outcomes. And while complexity helps us characterize more precisely the interaction between systems and the individual, the state of mind of the individual agent matters too.

Changing the Mind of the Agents

Wouldn't it be great if there were a simple pill one could take that would create a stronger affinity with the natural world, so that people's norms for caring about the climate would shift? And if this simple pill had no side effects? Surprisingly, it may exist. This is the astonishing age-old tale of a class of mind-altering substances. To some it may be a wacky tale, but one for which science is rapidly weaving an ever more solid foundation of evidence.

Harvard University plays a big role in this story. Richard Evans Schultes is considered the father of ethnobiology, the study of the relationship be-tween plants and people.[15] The son of a plumber, he studied at Harvard and became a professor there, spending decades exploring the Amazon forests and cultures in the 1930s and 1940s. He was the first to identify academically—and experience—the mind-altering substances that local tribes had distilled from nature and that they had used for eons. A Swiss chemist, Albert Hoffman, who accidentally rubbed one such substance in his eyes in 1943, synthesized an active ingredient similar to the one in many of these plants. Intrigued, he took a bigger amount on what he called his bicycle day, as it kicked in during his bike ride home. He called it LSD.

This was the dawn of the golden age of LSD and other psychedelics in the 1950s and 1960s.[16] Thousands of science papers were published on the miracle drugs. It started a deep research tradition into the functioning of the human mind, and owner Sandoz made LSD freely available to anyone with half a reason to research it. In 1957 an enthusiastic cover article appeared in *Life* magazine by JP Morgan financier Gordon Wasson. Alcoholics Anonymous cofounder Bill Wilson considered it a miracle cure for alcohol addiction. Futurist—and former Shell executive—Peter Schwartz questioned, "Why were engineers in particular so taken with psychedelics?"[17] Schwartz, himself trained as an aerospace engineer, thinks that it is because, unlike scientists, engineers work on problems involving irreducible complexity. "You're always balancing complex variables you can never get perfect, so you're desperately searching to find patterns. LSD shows you patterns."[18]

Enter Timothy Leary, another Harvard professor, who was to attempt to change the course of history. His colleague Schultes has little respect for Leary, notably his inability to get the Latin terms for the hallucinogenic plants right. But Leary had read the *Life* article and like some other scientists had become convinced of LSD's capacity to change society. Along with other evangelists, he wanted to change social norms toward closer harmony with nature. And end the war in Vietnam. And turn society away from mass consumption, toward the enjoyments of life that Keynes might have had in mind.

Inevitably, the reaction from a threatened social order came, in 1968. In a notable contrast with the current legal trade in opioids mentioned above, psychedelic substances are not known to be addictive.[19] With only rare negative side effects, and presenting substantial medical applications, they were nonetheless added to the list of forbidden drugs in the United States. In 1971 most psychedelics were also added to the UN list, sealing the prohibition globally. This action has had deep consequences, as it has painted a smorgasbord of different substances with same brush. Profoundly addictive substances like opioids were understandably included, but nonaddictive ones like psychedelics were also swept up. And deeply addictive substances like alcohol or nicotine were excluded. No consistency of policy.

Over recent decades, there has been a resurgence of scientific interest, although it has not been without controversy. In 2009 the home secretary of the UK asked Professor David Nutt to resign as chair of the Advisory Council on the Misuse of Drugs (ACMD) for saying publicly that psychedelics are far less harmful than alcohol or tobacco.[20] Nutt's university, Imperial College, had been at the forefront of renewed research on the effect of psychedelics. Nutt was merely quoting his recent publication in the medical journal *The Lancet*.[21] A topsy-turvy chart from the paper topples much of the common lore about drug impact, with alcohol and heroin at one end of the scale, and psychedelic mushrooms and LSD at the other. Apart from getting Nutt fired from the UK government, the chart has had a quite a run. In 2019 the Economist, as part of a plea to reconsider drug policy, reproduced it. It quantifies the collective impact of the individual and societal damage from each substance. The two worst offenders, alcohol and heroin are quite different in their impact. While alcohol's individual impact is smaller, its "harm to others" that of heroin. At the low end LSD and psychedelic mushrooms have low "harm to self" and no "harm to others." Of course even low harm is not nothing, but proportions matter.[22]

There is an intriguing connection to climate policy—the reason for our interest in the context of this book. A team at Imperial College has extensively tested volunteers on a standard psychological scale that measures nature relatedness, the response to statements like "I am not separate from nature but part of nature." Strikingly, a single psychedelic experience materially elevated people's scores.[23] As with all research, there are limitations to such measurements, but we can take solace in the fact that the conclusion is entirely consistent with the observations of scientists like Schultes who studied indigenous societies (who with a single exception all extensively use these natural substances from plants).[24] It is also consistent with observations by writers such as Aldous Huxley and many other individuals from John Lennon to Steve Jobs. Michael Pollan, through his extensively researched book *How to Change Your Mind*, is arguably responsible for the renaissance of the Western interest in psychedelics. Pollan observes that it is intriguing that the research shows that psychedelics are effective with two of the core problems of our time, namely, the environmental crisis and tribalism: people's affinity with nature increases, and the centrality of their own ego diminishes.[25] Although the mechanism for policy action is unclear, the importance and relevance are not. From a complex systems perspective we have been looking at ways to change the dynamics of the system. Any actions that also change the properties of the agents will make a climate revolution easier. Take for-benefit corporations. Consumers who have a greater affinity for nature and are less obsessed with their identity will more quickly express different consumer choices.

There would be an ironic elegance if an element of the solution to the climate crisis would be found in a substance that is ubiquitously present in plants around the world, and which have played a core role in cultural and spiritual traditions globally since the dawn of time.[26] The point is certainly not that we should add psychedelics to the drinking water in order to address climate change. That would obviously be wrong on many levels. However, reviewing the 1968 regulation of these substances, accompanied with appropriate safeguards, could open a new path to defining the social norms that form our connection to the natural world. Based on current scientific evidence, it seems a low-risk climate policy, with no known health downsides. We might just change our mind about our priorities for dealing with the climate crisis and—according to Peter Schwartz—better understand complexity as a side benefit.

Drugs are bad. Enterprise is good. Consumption drives economic wellbeing. Right? Not so fast. In this chapter we have attempted to unpack these all-too-facile ideas, throw them in the air, and reassemble them slightly differently. Some legal addictive drugs, like opioids, are an unmitigated disaster, driven by a market created from scratch in the pursuit of profit. Other nonaddictive drugs, like psychedelics, have helped form our connection with nature for millennia, and could plausibly do so again. At an institutional level, the very DNA of the corporation has driven it to manufacture an addiction to consumption. And there may well be better ways to treat our consumer addiction. What are they? New institutions, such as for-benefit corporations, could change our consumer addictions. Drugs are not drugs, consumption is addictive, companies are not companies. To foster a climate revolution, we need to go inside all these words and take a closer look. And recall that the challenge we set is to find solutions that scale to the problem, not blithely wishing that everyone did this or that.

No Single Recipe

This book counts on its readers being willing to put on hold deep psychological mechanisms at the service of understanding alternative policy options. But those who have come this far are probably tickled to do exactly that.

Recall that complex systems are composed of a large number of interacting components, without central control, whose emergent global behavior is more complex than can be explained or predicted from understanding the sum of the behavior of the individual components. In this chapter we have framed addiction to consumption as the emergent behavior, and we have asked how we might tweak the system so as to have more desirable emergent behavior—that is, more desirable social norms. To accomplish this, we might change the nature of the institutions that drive the emergence of consuming behavior. And we might influence individual consumers so that they become more responsive to these new institutions.

We have identified, somewhat speculatively, but not implausibly, that something like for-benefit corporations could do the trick and can easily be made available broadly, as we have seen in the United States. It is too early to tell whether they can help change social norms. While most of the

5,000 U.S. for-benefit companies are small, in 2017, for the first time, this kind of corporation went public: Laureate Education closed its initial public offering in February, raising $490 million.[27] The number of countries offering or considering the structure is rapidly increasing.

The renaissance in psychedelics is a potential avenue to modify individual consumers' preferences, given prudent regulatory changes. Such preferences undergo the kind of network propagation that social norms such as adopting solar panels or increasing obesity levels are subject to. That is necessary to allow for rapid scaling. These two are certainly not exhaustive measures, nor perhaps even the most doable, nor the most effective. And the solutions will be different in each culture and context. The intent is merely to show how a combination of tweaking the ecosystem design with tweaking the individual agents has the potential to lead to the required consumer revolution.

Camus asked, "Can man alone create his own values?"[28] Somewhat. While we think we make our own choices, the complexity lens suggests that those values are linked to the systems that surround us. Our current set of consumer norms is overwhelmingly an emergent property of the current norms factory. It is merely what we are used to. Reengineering the factory can unleash a different set of norms, which plausibly will feel just as comfortable.

If you buy a teapot at a flea market for a dollar and bring it home as your possession, and then your neighbor offers to buy it from you for his collection, you'll want more than a dollar as a fair price—even if you do not want to make a profit. Daniel Kahneman and Amos Tversky coined a term for this human trait: "loss aversion."[29] It gives more satisfaction not to lose a twenty-dollar bill than to find one. Admonitions to eat less meat or to take fewer flights are unlikely to be effective at the scale required—just like admonitions not to smoke. Equally, loss aversion makes it hard for us to imagine that a different set of social norms will be comfortable.[30]

It is hard for most people not to think of climate policy in terms of looming disaster or deep sacrifices. The narratives about climate change are mostly—albeit not entirely without justification—articulated in terms of doom and gloom. Although the IPCC has recruited armies of climate scientists and legions of economists, it omitted to engage psychologists and anthropologists—not to mention complexity scientists. Those social scientists might have helped shed light on how both individual and collective action actually happens.[31] They know that messages of doom do not lead people to take action. Anyone who has seen a smoker take his next cigarette from a pack covered in pictures of cancerous lungs realizes this. "Da Nile is not only a river in Egypt," according to a famous anonymous quip, and denial is a strong human tendency. Even faced with robust evidence of bad news, we seek refuge in alternative facts. Why should I change anything about my habits when the Chinese and the Americans are emitting most greenhouse gases? Why not spend the holidays in Hawaii if flying represents only a few percent of total emissions?

Whether the for-benefit corporation structure is the optimal institution or whether psychedelics are the optimal way of changing agent affinity for nature is not the point here. What matters is that fixing the climate crisis goes far beyond making more green electrons. It requires reflecting on what new institutions will act as ecostructures generating new emergent consumer norms and behaviors. And it also requires reflecting on what options we may have to change consumer behavior more directly. For-benefit corporations and psychedelics are just examples of such solutions, but the most effective ones can only be discovered through purposeful experimentation and innovation.

In turn, reengineering the social norms factory requires broadening the usual experts who advise on climate policy. Psychologists, political scientists, sociologists, anthropologists, and complexity scientists must take their rightful place beside climate scientists and economists. We should explore institutional changes that may help change consumer addiction, such as for-benefit corporations. We should consider reconnecting with age-old habits to change our connection to the natural world. Like all complex systems solutions, the recipes will be context dependent, but these suggestions illustrate that you can find practical steps to drive a revolution in social norms from the bottom up. Undoubtedly other ideas will be discovered along the way. And note that such measures come at very little cost—and are supported by research, which further limits risks.

10

Beyond Incumbent Industry

Change in the economy is driven more by the entry and exit
of firms than by the adaptation of individual companies.

ERIC BEINHOCKER

In 1997 BP's chief executive, Lord Browne, called for precautionary action
to cut greenhouse gas emissions. With this, BP was the first oil company to
publicly acknowledge the reality of climate change. At the time, it was
groundbreaking, and BP continues to be unambiguous about its policy: "We
believe that carbon pricing provides the right incentives for everyone—energy
producers and consumers alike—to play their part in reducing emissions."[1]

Except when it matters and BP isn't. During the 2018 midterm U.S. elec-
tions, Washington State put a carbon pricing policy on the ballot. It had
been rejected during a previous election and redesigned in consultation with
concerned parties. As with any policy, undoubtedly it was imperfect and
the result of uneasy compromises—but it was the first proposed comprehensive carbon tax on a state ballot in the United States. Proposition I-1631
would trailblaze a price for carbon pollution. It would start at a modest
$15 per ton in 2020 and rise by $2 a year (plus inflation) until 2035, when
it would reach a still-modest $55.[2] Still, adopting it had the potential to
start a network effect and accelerate carbon pricing across the continent.
Precedents matter. In the absence of active federal climate policy, contagion between the states is the only game in town.

But industry unleashed a veritable communications blitzkrieg to defeat
the measure—and scored a resounding victory.[3] BP was by far the largest
funder and lavished $13 million in its opposition, catalyzing total funding
of over $32 million.[4]

Such discrepancies between companies' corporate positions and their
practical actions are far from unusual. The decades-long Volkswagen diesel
fraud in the light of its clean diesel marketing is an extreme case, but incon-
sistency between aspirations and action is common. Dishonesty, cognitive

dissonance, or rogue local business units failing to follow the corporate party line? While that may happen occasionally, there may well be a deeper and more existential mechanism at play: deep path dependency and a lack of adaptive capacity on the part of companies. Some may simply be unable to execute the corporate sustainability vision, as they remain firmly rooted in corporate history. Companies are primed for efficiency, but this comes at a price. Their resilience and adaptive capacity are vastly diminished. Incumbent companies just don't do revolutions.

Blockbuster had a great idea—it provided video rentals to people and was itself a blockbuster.[5] The key technology that made that possible was database analysis that gave them the power to analyze demographic characteristics of neighborhoods to decide what movies were most profitable to stock. With Blockbuster, people could go to the store, pick out the movie they wanted to see, and watch it in the comfort of their home, rather than wait for it to be shown on TV or go to the theater.

In the late 1980s, that was a formula for success, but by the 2000s it was already being challenged. The problem was that the formula was designed around in-store pickup, and with the technology change to DVDs, mailing became an option, which meant that Netflix grew and online transfer was becoming the far more efficient way. To succeed, Blockbuster would have to switch to these other methods. The problem was that its employees weren't versed in those other methods, and Blockbuster was always a year or two behind. But they spent billions of dollars trying to switch, which undermined the business they had. In 2010 they went bankrupt. Curiously, in 2019 there is one single store left, proudly bucking the trend in picturesque Bend, Oregon, otherwise known for its cannabis dispensaries and craft breweries.[6] What would have been the appropriate thing to do? Stockholders would have been a lot richer if Blockbuster had simply decided to grow old gracefully, and to give franchisees more flexibility to modify their own businesses into the newer technologies if they wanted to do so. Successful ones could have expanded and developed and evolved. That didn't happen. Nor did Blockbuster fold deliberately. It could have, at that point, simply wound down slowly, preparing financially for the inevitable. Instead of bankruptcy, it could have gone into a pleasant assisted-living and eventually well-off death. As unlikely as it now seems, Netflix plausibly awaits a similar fate.

Invisible Networks That Bind

Businesses develop an internal culture and set of interconnections that adapt well to one particular set of skills. They create lock-ins, deliberately. Many companies think that the structure on the management charts provides the map of the inner workings of the company. That isn't so. In every company there is an invisible management system that makes the formal structure work. That invisible structure smooths out the rough edges and solves problems that need solving—it creates its own hierarchy that works for the particular problem. When the problems and the institutions change, even if the formal structure is changed, the informal structure doesn't adapt nearly as fast. That creates tensions.

In biological terms, the informal structures are like the nerves, which take months to grow only a bit. Thus, in a transplant one can connect the veins and arteries, muscles, and bones relatively quickly, but the nerves are much smaller and need to grow on their own. And that will only happen if the environment is right.

It is these informal networks that make it possible for real-world institutions to operate efficiently. But it is the same invisible structures that also make it difficult, if not impossible, for large organizations to adapt to changing conditions.

These informal networks that make systems work create a path dependency, which is nothing more than saying that they are rooted in a past that gets in the way of the future. There are ways of making these informal networks visible. One imperfect way of getting a sense of this network might be through plotting a map of communications and connections, from social network patterns and e-mail traffic inside the company.[7] There's even software now that will help you do that. You would then notice that the patterns of communications don't follow the lines of the charts—but that there are all sorts of surprising nodes of connectivity. It could be someone running a football betting pool, but more likely you'll uncover something of the invisible management system.

Take AT&T—until the 1990s it was the largest telecommunications company in the world, with a deep technical legacy epitomized by the multiple Nobel Prizes collected over time through its research arm, Bell Laboratories. It also pioneered human factors design and behavioral studies. AT&T was also at the cutting edge of management disciplines, initiating or adopting quickly new ideas—as well as fads. In 1984 AT&T had just been broken up into eight "Baby Bells" and the remaining entity, and in exchange for losing its local monopoly in the United States, it was now no longer excluded from operating abroad, as it had been for the previous sixty years. In brief, it was a reflective company, set on global expansion. Yet one of the things that was immediately noticeable when I joined the company was that its cost accounting was incredibly opaque.[8] Understanding how much products and services actually cost the company was a painful exercise of digging through reports and definitions, and you were never quite sure whether you had discovered everything.

After asking around and grumbling about this, an older senior finance person once gave this plausible explanation: Throughout its history, AT&T had been obliged to justify its cost to the regulators in order get it rates approved. In such an environment more transparency would have led to debates and increased the likelihood of rate cuts. Over time, subtly but inexorably, this led to a set of systems and a culture where costs would be spread, reallocated, and distributed in such a way that no one quite knew where they had ended up. It was a bit like one those sidewalk scams where a deft handler shifts a pebble between cups, taunting passersby to guess under which cup it ended. There was nothing devious or evil about this; rather, it is simply the emergent behavior of a system encouraged for a long time in a given direction. The result many years later, however, hampered AT&T's international expansion, through a dependence on a historic path.[9]

If companies are hamstrung by their past, how can they sustain a competitive advantage and innovate to stay ahead? Complexity suggests that the crisp answer is that they can't. Eric Beinhocker articulates this in the following way in *The Origin of Wealth*: "The virtual nonexistence of excellence that lasts multiple decades, and the extreme rarity of repeated excellence, brings us to a brutal truth about most companies. Markets are highly dynamic, but the vast majority of companies are not."[10] Companies can innovate, and they do, but new companies and markets are just much better at it, as they don't have the baggage existing companies have.[11]

Resiliency versus Efficiency

Complexity suggests that there is an additional reason that companies are poor innovators: in the trade-off between resilience and efficiency, most companies come down squarely on the side of efficiency. Competitors, markets, and shareholders all drive them to be as efficient as possible to excel at their core business activities. They don't have a lot of choice in this respect, as it just reflects the institutions that have been created to host them. It could have been different, but it wasn't. Inevitably this sacrifices some or all the capacity to be adaptive. Nature provides an illustration of this. If you visit the panda reserve in Chengdu, China, you are struck by the panda bears who are supremely efficient at eating fresh young bamboo shots, but pretty useless at anything else, except looking cuddly. This includes struggling with reproduction from a very lazy sex drive. This is in sharp contrast to the swarms of beetles scuttling around them, with an innate ability to adapt to different kinds of food, to find new routes and take over tasks from each other. Similarly, companies typically excel at very few things, as a result of overspecialization. High efficiency mostly comes at the price of low resilience.[12]

Since the creation of the East India Company in 1602, publicly traded companies have been the most visible part of the market economy. But many companies are either private or state-owned. While the trade-off applies universally, the balance swings most strongly toward efficiency for publicly traded companies. Private and state-owned companies are more insulated from the forces of the market and as a consequence have more freedom to invest in adaptive capacity. Adaptive capacity shouldn't be confused with inefficiency—and lack of market pressure can easily lead to inefficiency. The point here is that efficiency is not free, and while our argument focuses on publicly traded companies, the fundamental logic holds for other forms, although their pressures will be different.

Locked In before You Know It

The early success of Tesla illustrates how incumbent car companies are unlikely to succeed at electric cars. Sure, it may be too early to call this particular race, as every other car company announces ambitious plans to make electric cars. But based on everything we have seen before and described above, they are likely to struggle with numerous soft barriers. The different engineering approach with an order of magnitude fewer parts than internal combustion engines, their historic supply chain relationships, the network of dealers that depend on repairs and maintenance—all conspire to make it implausible that most of them can catch up with Tesla. The stock market appears to know that too.

However, for all its remarkable innovation, it is also possible that Tesla is already locked into a model of individual car ownership. The revolution in mobility we'll look at in Chapter 12, one of collectively owned super-lightweight autonomous vehicles, may plausibly already be out of reach for Tesla. Time will tell.

Companies are locked into more than their own business model; they are also locked into the larger system of consumption. As we have seen in Chapter 9, the design of the institutions and policy frameworks conspire to increase our addiction to ever-increasing consumerism, well past the point of optimal utility—or even pleasure. Yet just as one can't blame fish for the ocean being wet, one can't blame companies for consumerism. It's the ecosystem they are in that makes them do it. In dealing with the climate crisis there is a great deal of finger-pointing and demonization, including of companies. According to the Carbon Disclosure Project, "Since 1988, a mere 100 companies have been responsible for 71 percent of the entire world's industrial greenhouse gas emissions."[13] Convenient, but misleading. The implication is that if I leave the heating on and the windows open, the ensuing carbon emissions are the fault of the power company. Absurdly reductionist. Looking at the economic system as a complex system, it becomes clear that the behavior of companies is consistent with the design of the economic system. The economy is not a natural system, but one with man-made institutions and choices. Just as individuals' norms coevolve with their environment, so do companies.

This doesn't mean that corporate responsibility is a hopeless idea. Companies should be pushed to act as responsibly as possible, within the ocean they swim in. There are enormous improvements that can be made within the status quo. Reducing waste, designing for recyclability, energy efficiency, using renewable power, and implementing biodiversity policies are all areas where business could and should make valuable contributions. PWC's Chief Purpose Officer Shannon Schulyer puts it this way: "It's how you connect purpose to an individual so they know what they need to do in their roles and how do you help them see personally how they connect with values and behaviors."[14] The thinly veiled purpose is to keep staff on board within the current business model, rather than exploring what the broader context requires.

Incumbent industry is an engine of societal stability and wealth creation. It serves the equilibrium state and indeed deepens it. To that end it favors efficiency over resilience. That is good and useful. Except when the system needs changing, when regulations are needed. Then it's not so useful, and it requires a refresh. New businesses emerge to serve and indeed shape the new order. That is how it has been in previous industrial revolutions.[15] That is how it will likely be again. And that is not just an empirical fact; it is also intelligible from the very nature of complex systems.

The contribution of the incumbents is likely to fall well short of what is required to deal with the climate crisis. The economic system needs to be tweaked so as to lead to different emergent behavior companies. This is the responsibility of policy makers—albeit with the strong support of CEOs, not in their day jobs, but after hours, as members of society endowed with privilege and responsibility.

Adaptation Is for the Young

It is hard for a company to adapt, and one of the most important lessons from complexity is that often the best strategy is to die gracefully, since the innovation outside firms is almost always faster and more original than inside firms. One way for a company to die gracefully is to hand back money to the shareholders, effectively recognizing that it has nothing better to do with the capital. It could raise more money to hand back or for winding down by selling those pieces of the company that have value for others. The reason a company is successful is often mastery over competitors in a very specific area in which it is highly efficient. But when technology changes, that reason can become a handicap, and it doesn't stand much chance of succeeding in the new environment. But it can continue to do what it does best in a shrinking market, and enjoy dying—paying its stockholders dividends rather than wasting those dividends on an implausible transformation.

Firms are mostly good at executing complicated tasks, not complex ones. (Conglomerates such as GE and Tata are possibly rare examples that manage to get the best of both worlds, for a while at least.) But the pain from downsizing and failure of firms has both an individual and a social cost. Arie de Geus argues in *The Living Company* that firms need to learn, adapt, and survive in order to avoid the pain of when they fail.[16] He refers to this as "planning as learning." While this is important for short-term performance, it is unlikely to allow the company to dodge corporate demise.

Of course firms should work very hard to survive and adapt, but society should leverage a Darwinistic cycle of boom and bust for its collective progress. This in turn implies that it is both necessary and efficient to put in place appropriate tools and social systems to absorb and facilitate the transfer and retraining of employees to other companies. And if you are still counting on a job for life at your company, realize that the probability is very high that you will outlive your employer. For society, dying is not an option. A hurricane didn't erase New Orleans, but it only took a couple of creative MBAs to destroy Enron. Hiroshima is a flourishing city with a traumatized past, but those old enough to know the brand may wonder where Datsun is today. The point is that cities are much more resilient than companies.[17] A realistic assessment of a company's strengths and weaknesses and an avoidance of exceptionalist views makes fading away easier to envisage.

There appears to be one recipe that provides a way out of the lock-in, to a new future. Notably, it involves a mini-revolution inside the company itself, as it divests enough of its core business—and associated identity—to open the possibility of reinventing itself. The examples are few and far between. IBM is one of the rare companies to survive the IT revolution, buoyed by selling its core personal computer business to Lenovo. After sitting on the sideline of the Energiewende, German utilities RWE and E. ON, which we met in Chapter 7, unshackled their business into a legacy and a new-energy business. The former Dutch state coal mining company DSM transformed itself into a nutrition company by aggressively divesting all its former fossil fuel operations. But these are rare exceptions to the strict rule of efficiency over adaptability.

These patterns repeat themselves again and again, even in such brand-new markets as electric planes, where Airbus has already stumbled with its first model,[18] and EasyJet is partnering with start-ups like Wright Electric to realize its vision of electric flight for all trips under two hours by 2030.[19] Again too early to tell, but the pattern remains consistent.

While this applies to much of industry, there will be considerable variation per sector. The position of the service sector may be less tenuous. In the financial sector, intermediaries such as real estate brokers may find it easier to adapt to the revolution. They exist to serve their clients' needs, so they may find it easier to change accordingly. Some construction companies may find ease in building zero-energy buildings, while some architects maybe stuck in their ways—or the other way around. In any case, the incumbent industry will plausibly have a limited role to play in any revolution. The people in them, as well as their processes and ideas, will provide crucial contributions as they join new initiatives and start-ups—but in their current incarnations they are unlikely to be very relevant. Shareholders introduce resolutions all the time for oil and gas companies to become renewable companies. That is most likely a waste of effort, as these incumbents are unlikely to contribute meaningfully to the energy revolution.

Society needs revolutions, but the existing companies are not the obvious partners—with exceptions. That is not a lack of capacity, talent, or imagination, but merely the consequence of a structure that is designed to optimize for efficiency. The corporation is an institution that was designed and refined over time, with the specific intent to share risk, manage liability, and be efficient. Almost inevitably, efficiency comes at the price of adaptability. Mariana Mazzucato argues in the *Entrepreneurial State* that fundamental innovation overwhelmingly comes from the state sector, for example, from universities, research institutes, and the military.[20] It is business that implements these ideas with great focus and energy, reducing costs through learning by doing and subsequently creating markets at scale.

For revolutions through policy, the incumbent industry is likely to simply be too strongly constrained by its own path dependencies. This represents a real dilemma, as those companies have strong positional power, and by definition the new companies that are likely to deliver the climate solutions do not. There is no easy fix for this, but it starts with a realization of the relatively minor role that most incumbent companies will play in resolving the climate crisis. There will be exceptions such as IBM as a rare survivor of the IT revolution. Companies are not to blame; their institutional design was intended to value efficiency over resilience and adaptability.

Under normal circumstances, it might not be necessary to be quite so undiplomatic about the nature of business and to call this out so bluntly. Having spent much my career in business, it even slightly pains me to do so. But these are not normal circumstances. In order to deal with the climate crisis, we need to be very clearheaded as to the nature and the capabilities of the pieces on the chessboard, so that we can use them accordingly. Companies can do a lot better: Unilever championing the integration of sustainable development goals (SDGs) into business strategy, and—yes—BP acknowledging climate change in 1997 are both important developments. But incumbent companies can by and large not be agents of systemic change. Policy makers should recognize this and not make the survival of the incumbents a cornerstone of their climate policies. The incumbents certainly matter for the health of the economy, which in turn underpins the societal willingness to accommodate changes. But for the change itself, they will generally be less helpful.

Business matters, but to address the climate crisis it's mostly new business. In those new companies, people will apply lessons from their previous employment. People are resilient and adapt; businesses for the most part do not. The consequence for policy makers is that they should be less concerned with the interests of incumbent industry and more with the capacity to foster new ones. The incumbents will have smoother talk and meatier campaign donations, but it is the start-ups of today that will deliver the revolution of tomorrow.

Who Pays for Learning?

> [Obama] put money into a whole mess of companies. About
> 90 billion dollars into green energy companies like Solyndra
> and Tesla.
>
> MITT ROMNEY, 2012 CANDIDATE FOR U.S. PRESIDENT

Picking Up Twenty-Dollar Bills

Picking winners is hard, which is why the performance of venture capitalists (VCs) is so challenging.[1] Many try and fail to pick the next Amazon, Facebook, or Uber. In fact, most companies picked by even the most successful venture capitalists fail. That's not to say venture capital's services aren't important. They are. There's a reason for the name, though: venturing into new territory.

It's also clear that venture capital and government don't quite mix. We have alluded to the Solyndra bankruptcy, but Solyndra is only one cautionary tale. Talk to many a venture capitalist up and down Silicon Valley, and you hear people questioning why the government chose Solyndra to support, sometimes followed by a derisive "In fact, we passed. That's why they had to go to the Feds." After the fact, that's a rather easy statement to make, but there's something to this argument: had there been money to be made, smart private investors would surely have jumped at the chance. Standard economics is beset with truisms like that. There are no twenty-dollar bills on the ground, the saying goes, because if there were, someone would have picked them up already.

That statement says both a lot about economic logic—what makes economics so logical—and why and when it fails. If there were obvious, free twenty-dollar bills lying on the ground, there would be easy money to make. That is clearly true at certain times in certain locations. Think gold rush, or its modern-day equivalents—whenever a new location (Russia, China, Iraq, Iran) or a new field (biotech, high tech, fin tech, green tech) opens up. Even in all those cases, there are some winners and a great many losers.

There's also much more to picking winners for a VC than first meets the eye. Picking winners is about more than just money. Well beyond providing liquidity to a struggling startup, there is the additional warm glow of being one of the anointed, chosen by one of the Silicon Valley blue chip VCs to be the Next Big Thing—much like an equity investment from Warren Buffett's Berkshire Hathaway means more than just money for staid corporations. It comes with that extra oomph, and a guaranteed spot in the news. VCs, of course, know that.

There's also something to the standard economic logic, though. While the going is good, others sweep in, until there are so many competitors squeezing out each last ounce of efficiency to make entry no longer worth the cost. Marginal costs and marginal benefits will be in equilibrium, to use a favorite economic term. In equilibrium, then, there can't be any such twenty-dollar bills lying on the ground. As discussed in Chapter 4, equilibrium, of course, is a funny concept. It's unachievable in the short run, and "in the long run, we are all dead" anyway, as one of the most famous, now-dead economists once said.[2] It is, thus, a poor guide for policy, and a poorer guide for business. The task is to find the right opportunities where available and to stamp out inefficiencies where necessary.

Back to Solyndra. The company name has become synonymous with anything from government overreach to crony capitalism, to green investment folly, to any narrower political angle one wants to attach to the matter.[3] The fact is that Solyndra failed. It did so spectacularly, taking around $530 million of public money with it, as it filed for bankruptcy two years after a quick loan approval process that could have been the poster child of efficient government at its best had it succeeded.

It was not meant to be. The immediate cause of Solyndra's demise had little to do with a failing clean tech sector. If anything, it was the opposite. Silicon solar photovoltaic (PV) panels became so cheap so quickly that Solyndra's supposedly more advanced thin-film technology had no chance of competing in the marketplace.[4] That's the cycle of innovation at work. Some win, some lose. Sometimes the right technologies win and set the new standard for generations to follow; sometimes they don't.

Overall, the Department of Energy's renewable energy loan program made about $5 billion on loans totaling a bit under $35 billion.[5] Those returns are better than many VCs achieve. What's more surprising: those returns came despite the fact that, unlike with most VCs, the profit motive alone was not the impetus for setting up the program in the first place. It was a public program, after all. And it was put in place in the depths of the financial crisis, when it was clear that private markets had dried up and that a public stimulus could help fill significant short-term financing gaps.

If anything, the program's failure rate was too low, pointing to an inappropriately cautious approach on the part of the government. It could have—should have—taken more risks with public money, possibly for even greater net benefits of the program. There are any number of innovations that need support to increase energy efficiency in the built environment—triple glazing that is cheap to install, techniques to insulate walls in existing buildings, household heat exchangers. Or more research into batteries. Or reinventing the rampant air conditioners that are dueling rising temperatures.

One of these risks was Tesla. Only Elon Musk will know the full story behind its $465 million loan, why Tesla went to the government for money in the first place (a good guess: cheap, easy money), whether it could have made do without the extra money (probably), and how it paid back its loan nine years ahead of schedule (finding twenty-dollar bills at the right time and place). Here too the glass-half-full verdict could be that the government took too little risk, or at least it handed out its loan too cheaply. Slightly tougher loan conditions could have made for a slightly better return for the government. Of course, there is no way of knowing before the fact, which is the whole point of investment risk. But certainty clearly wasn't the point. Government, after all, is here to serve the public good. If its loan decisions made some investors even richer, so be it. That's a small price to pay for jump-starting what has since proven to be more than simply a niche market. Tesla itself may not end up owning the electric car future, but it clearly helped spark a revolution. Several other traditional car companies are now rushing in. Mercedes and BMW colloquially refer to their own electric car designs as Tesla-Jäger—or Tesla-chasers. This is a revolution in the making, and the government loan contributed.

The loan program, of course, is more than headline companies. It also helped finance the first commercial-scale solar PV plants in the United States. Would those have happened without the government? Perhaps. Would they have happened as quickly? No.

In this context, the failed Solyndra investment should be seen through the lens of a broader portfolio strategy. While focused on commercializing solar PV, the Solyndra bet was a small hedge.[6] One or the other had to fail. With traditional PV succeeding, Solyndra failed. Had Solyndra succeeded, at least some of the traditional PV investments might not have. Fortunately, they did, in part thanks to other government programs that supported the initial deployment of traditional PV. Direct payments, price guarantees, loans, or other. The government, as lender of last resort and insurer of risks that traditional VCs would not take, provided a public service.

Did it provide the right public service?

Investing in Learning by Doing

Innovation—even, or perhaps especially, failed innovation—provides a public good. It creates knowledge for those who come after. Standard economics all but demands a public investment to provide more innovation than would otherwise occur. That's not to say, of course, that innovation wouldn't happen without subsidy, but free money surely helps.

Innovation happens on several levels. There's fundamental research and development. Basic science is a clear public good. Particle physics, fundamental math, even sequencing DNA ought to be supported by public funds. Some of the money will be wasted. Some will lead to fundamental breakthroughs—from quantum computing to extending human life spans. Some of this research will happen in universities, some in dedicated, well-funded research institutions, some in the military.[7]

DARPA is well known for pushing the frontiers of science.[8] Some of the inventions have a direct, immediate military application. Many don't. Though one thing DARPA has recognized from the very beginning is that R&D alone is not enough. The acronym merits at least another D for "deployment." That applies for the military as well as for energy technologies.

Extracting the first drop of oil was hard. Few even knew what to look for, and why. It took years for John D. Rockefeller to be hailed as a hero and to become the founder of an entirely new industry.[9] Extraction costs quickly came down. The same goes for most new technologies. With the right approach, as we saw with World War II bombers in Chapter 2, new technology tends to follow a learning curve. That learning, more often than not, happens through doing: learning by doing.

California governor Arnold Schwarzenegger signed into law the Million Solar Roofs initiative. It was quickly renamed the California Solar Initiative, after it was evident that a million roofs would be out of reach. The naming snafu notwithstanding, the initiative was a success by the only metric that should matter: Did California get the size of the public investment right vis-à-vis maximizing net benefits to Californians?

It did.[10] And it followed a number of sound principles, namely, that the contribution ought to start high and decrease gradually almost immediately. Just so it doesn't distort markets permanently. Also, the subsidy ought to be focused on actual learning by doing. And subsidizing learning by doing is no replacement for pricing carbon. The last bit is important. It is tempting to conclude—à la Google's now retired "$RE < C$" equation—that renewable energy ought to cost less than coal—that it doesn't matter whether one prices CO_2, raising the cost of coal and other fossil energy sources, or whether one subsidizes renewable energy. In fact, the difference does matter because of path dependency. Renewables being cheaper than coal may not be enough to dislodge the latter from its equilibrium perch. Which suggests carbon prices should be well above the social cost of carbon, at least in the beginning.

Picking winners does indeed come with pitfalls, but we can be certain about picking the main loser: CO_2 and other greenhouse gases.[11] That's what makes charging a price for CO_2 so simple. It's economically sound, but it is not enough—necessary, but not sufficient—and potentially difficult. Carbon taxes are regressive, so those with lower incomes will pay the bulk of them. This will make it hard to get social acceptability. Just ask the Gillets Jaunes, with their iconic yellow safety vests, who erupted in widespread protest in France in 2018, rising against the increase in fuel prices that had just been proposed by the government. These price hikes seemed reasonable—until the climate crisis met the inequality crisis. Lower-income citizens, who had been driven out of cities toward distant suburbs by high rents, depended on long car commutes to work. French citizens in the third income decile (pretty much the folks who ended up protesting) account, on average, for about 6.5 tons of CO_2 annually. In contrast, individuals in the top 1 percent of the U.S. population churn out 320 tons of CO_2 each.[12] In the practice of the political economy, inequality and climate are inexorably tied.

The other side is equally clear. Learning-by-doing subsidies have a rock-solid foundation.[13] Think of CO_2 as a negative spillover. Its use ought to be priced to account for the full costs. Think of learning as a positive spillover. It ought to be subsidized. Learning by doing, in short, can only happen if someone does it. Without subsidies, too little will be done and too little will be learned.

"Subsidy" is an ambiguous term. It has two very different meanings. It can refer to public funds to contribute to an unprofitable product or service, such as nuclear power, and the subsidy then basically goes on forever. Or it can mean strategic temporary contribution so that enough learning by doing kicks in to make a product or service able to stand commercially on its own legs.

Government picking winners has its downsides, such as in the case of Solyndra. Even when it does pick successfully, it may not add much beyond cheap money for those who may not have needed it, such as Tesla.

Government money comes with other costs. For one, regardless of one's political suasion, it's clear that it is not actually free. Eventually, every dollar spent requires a dollar raised (although for some governments, such as Germany's, the cost of borrowing is even negative). And those taxes come with deadweight losses, creating additional inefficiencies in the economy.

Ultimately, society pays for the subsidies given out on its behalf. The added transaction costs, bureaucratically predetermined direction of innovation, and possibly diverted funds surely come with real costs to society.

Subsidies also have a way of sticking around well beyond their useful shelf life. Witness the hundreds of billions of dollars' worth of fossil fuel subsidies still in place to date.[14] Every one of these dollars is a step backward for the planet. Politically, it seems impossible to phase them out[15]—to avoid subsidizing the average ton of carbon dioxide emitted to the tune of $15.[16]

Subsidies are messy. Or at least they can be. They are indeed optimal, though, if they are focused on the learning-by-doing aspect, as those given under the California Solar Initiative are.

Solar PV provides a good example of how unexpected learning can be. Solar PV panel costs have plummeted a good 80 percent in less than a decade.[17] A good part of the drop can be attributed to supply subsidies, primarily in China, and demand subsidies, led by Germany. The combination pushed more panels onto the market than would have otherwise happened.

Was that rapid deployment efficient? In a narrow sense, probably not. Chinese households subsidized their domestic solar industry well beyond what was in their (short-term) interest. The same goes for Germany. Its Energiewende, which included heavy subsidies for solar PV deployment, increased the price of energy for German households. It led to rapid deployment of solar PV. Regardless of the direct implications for Germany's energy mix, and the complex interactions with European energy and climate policies, one lesson was clear: German subsidies caused solar PV to climb up the learning curve and, thus, slide down the cost curve quicker than would have otherwise happened.

Much like subsidies under the California Solar Initiative, German feed-in tariffs have fallen ever since they were introduced. A cynical observer might conclude that Germany was simply running out of money. But much of the tapering off of subsidies was by design, taking a page out of the learning-by-doing book: initially, feed-in tariffs ought to be high, declining over time. Germany, like California, followed that fundamental pattern.

The Road to Cheap PV

Headlines shout it from the rooftops: "Solar panels are cheap and getting cheaper"; "the dramatic drop in the cost of solar PV modules, which has fallen by 99 percent over the last four decades"; "solar PV has undergone an industry-wide revolution." Even the IEA has joined the chorus, as every year before, adjusting their previous year's optimistic forecast upward by a third—and missing the irony of that action. It has become a truism that solar costs have fallen quickly. But why have they fallen? Sure, early American R&D, German consumer gifts, Chinese strategic industrial policy, contagion effects . . . The publicity helped when Jimmy Carter put solar PV on the White House roof in 1979 after the Arab oil embargo (and Ronald Reagan removed them in 1981 and Barack Obama put them back in 2010 and notably Donald Trump did *not* remove them). To purposefully realize the kind of cost revolution that characterized solar PV, it would be better to know in which proportions all these factors contributed.

Scientists from MIT and the Santa Fe Institute decided to find out. They built a model with all the various effects that contributed to plummeting costs, and proceeded to fit it to the actual cost reductions and the deployment volume. The deployed volume, after all, impacts the degree of learning by doing that only occurs when you build lots, roughly proportionally to all that has been deployed before. Unhelpfully for those who like simple headlines, but familiar to the complexity literati, some effects played a strong role for a while, and then others took over. Different factors wove in and out. For instance, improving efficiency was the largest contributor in the first period, responsible for 24 percent of the cost reduction. In the second period, module efficiency was only the fourth most significant factor, and its contribution dropped to 12 percent. R&D played a dominant role in the first period, improving multiple cost equation variables that reduced cost significantly, but its impact is high in the second period as well. Scale economies went from being a minor contributor in the first period to a significant one in the second.[18] The point is that different factors are at work at different times; it can't be explained with a one-liner. Of course, 20/20 hindsight is great, even if it is from complexity scientists, but it is even more useful if you can apply these insights to future developments. And this is exactly what this model allows. Policy makers will be able to better judge what elements are most useful at which time, so as to create purposeful revolutions.

Learning Is Faster Than They Think

Learning very often happens much faster than expected. The cost reduction of solar PV is much discussed, but not unique. Take environmental regulations that force companies to change products or processes. A commonly heard complaint is that this makes companies less competitive and their products more expensive. The amount of money required to comply with the new rule imposes a real burden on the consumer as prices go up. But that appears to be greatly exaggerated. When you compare the estimate ahead of the issuance of the new rule and the cost in hindsight, it becomes clear that most of the time the estimate was too high—and sometimes by a lot. Overestimations of between two and eight times are not rare—although underestimations occur on occasion.[19]

There are many such examples. The total annual costs associated with early proposals to control U.S. acid rain emissions were estimated at $3.5 billion to $7.5 billion. In hindsight, costs have been significantly lower, at under $1.5 billion.[20] But costs are not always vastly overestimated. The Netherlands in 1989 adopted a suite of 400 environmental rules. Looking back twelve years later, the initial cost estimates for implementing them had only fallen short by 13 percent, although this was partially explained by the widespread availability of lots of data over the period before the rules were enacted. And although talk about the high cost of environmental regulation rolls off the tongue of many a politician, an EU report summarizes that "overestimation of ex-ante costs is common, though not a universal rule."[21]

It's not that people are dishonest about the amount of effort it will take, although at times some excessive lobbying fervor might occur. More often than not, technical innovation exceeds people's expectations.[22] Learning by doing tends to be faster than our ability to forecast it. As described in Chapter 3, this is consistent with the complexity perspective on innovation, as an emergent process through the directed recombination of ideas.[23]

The implication is that the level of contribution toward paying for learning curves—we'll also call them subsidies for now, per the conventional use of the word—tends to be overestimated. So what is needed is a system that can do two things: flexibly support movement up the learning curve for a given tech, but apply it to a broad portfolio given the inherent uncertainties. The policy makers in the tech funding business, like DARPA, know this, but they need their political masters and the press to take industry pushback on regulatory cost with a grain of salt.

Subsidies ≠ Subsidies

Crucially, learning-by-doing subsidies are anything but what one colloquially perceives as subsidies, that is, subsidies to make up for a lasting shortfall; subsidies that will be around forever. Perhaps better to do away with the "subsidy" label altogether and simply refer to them as learning curve investment. LCI anyone? In any case we should remember that the same word gets used for two totally different meanings. Someone has to pay for the learning curve. Occasionally, the market alone will do it. Increased deployment in a relatively competitive market will lead to prices coming down over time. Sometime, it takes government help. Most importantly, some government help will be efficient.

That's true when only considering narrow economic benefits. It's doubly true when looking at broader spillovers like network effects. Solar panels, once deployed in a particular location, motivate others in the same town to do the same. And they likely influence other connected behaviors such as recycling. That effect is particularly pronounced in smaller towns, presumably because the community there is more tightly knit.[24] If I don't just live next to my neighbor but actually like her, I might be more willing to do as she does. But only if the government has provided the right dosage of LCI or subsidies for learning by doing—at least while it's required. The LCI will enable deployment scale-up, which in turn can drive an evolution of social norms with spillovers into other areas.

In short, even by standard economic logic, there is a strong case for government support in the learning-by-doing phase. The case is even stronger in the more realistic complexity frame. Government is needed to enable such pathways. And this is a role government has frequently successfully played.

Autonomous Cars as a Climate Policy

> In general, the science of complex systems is not about making exact and detailed predictions about specific events in the future. That is the business of prophets.
>
> STEFAN THURNER

In 1900 throngs of pedestrians made their way along the broad sidewalks of New York's Fifth Avenue; on the road were long rows of two- and four-wheeled carriages, each powered by one or two horses. In a 1913 photograph, all the horses are gone—replaced entirely by motorcars, and flanked by the same pedestrian-laden sidewalks. That's how fast a revolution can play out. Today, autonomous cars seem like a far-off dream for many—or a nightmare for some. But potentially it's another revolution. We can't be sure, and since I'm not a prophet, I won't make predictions.

For our exploration of new pathways to solve the climate crisis, autonomous cars are of interest for two reasons. They represent a potential revolution, so we can look at how revolution policies are different from standard policies, in this case in the absence of the benefit of hindsight. The second reason is that it's about much more than transport. Transport is deeply interconnected with other sectors, such as energy and the nature of the urban fabric. A revolution in transport could also catalyze revolutions in social norms. Freeing up parking space in cities could spark a revolution in urban design, as the opportunity arises to reconsider large swathes of urban space. And since we're not making much progress on climate mitigation, a revolution in an interconnected sector just might help dislodge other adjoining systems from their own locked-in states. Even if you don't care much for cars, autonomous vehicles are a fitting illustration of revolution policy at work. They represent a potentially radical technology shift that is at the same time familiar and bewilderingly different. The disruption is technological in nature, but the consequences are social and environmental—and the hard choices fall to policy makers. The attraction of autonomous cars as an example is that their future is relatively open, but tantalizingly close. Below we look at the dynamics that drive toward an incremental or a revolutionary path.

Good-bye Steering Wheel

The original Tesla S rolling off the factory floor sparked quite some excitement. But it really comes into its own when it leaves city boundaries and cruises along a highway, when the driver activates its autopilot and an algorithm takes over control. Fourteen sensors continuously scan the tarmac in all directions, the output of their readings represented on a sweeping seventeen-inch screen. The car adjusts velocity and tweaks its direction accordingly. Cameras watch out for speed limit signs and track the lines painted on the surface. When the road curves, the car adjusts its speed. The data from all this, and about any human intervention, is streamed back to Tesla HQ for continuous evaluation. The aggregated lessons lead to regular software updates—via automatic downloads—that help evolve the driving algorithms.

This kind of driver assistance is admittedly great progress over the previous "technology" involving human drivers gluing their eyes on the road and their hands on the steering wheel, with markedly mixed success. Drive a Tesla and arrive remarkably more relaxed. But in some sense this is just an evolution from previous improvements in automobile technology, such as airbags, or catalytic converters to help rid car exhausts of dangerous pollutants. The car is still a very heavy, individually owned machine, which the average owner uses less than 5 percent of the time.[1] And while in motion, the all-electric Tesla consumes roughly as much energy as any of its predecessors. During the other 95 percent of the time this particular Tesla is parked on twenty square meters of public land in the center of Amsterdam, enjoying a lavish and largely implicit public subsidy in the process![2]

The main reason this car—like most cars that have come before it—is so heavy is that humans crash a lot. As a consequence the car requires heavy reinforcements to protect the passengers from their own helplessness. Still, 26,000 people are killed in traffic accidents every year in Europe alone,[3] and over a million globally.[4] Traffic deaths are now the leading global cause of deaths for children and young adults between five and twenty-nine years of age.[5] Many more are injured, although this number has been reduced through clever armor design. The Tesla packs more mass still, in the form of 600 kilograms of lithium batteries. The battery is so big and heavy in good part because of all the heavy safety features of the car. Such feedback spirals all conspire to make it heavier still.

Meanwhile, the weight is the main reason for the massive energy consumption needed to propel it forward. As marvelous as the driver-assist features appear, they are in reality just an evolutionary and incremental improvement. The design is stuck in a vicious circle of accident-prone drivers, who require heavy cars, which in turn requires heavier engines. The inability of humans to drive safely leads to heavy and large electric batteries.[6] If people didn't have so many accidents, cars could be much lighter and drive much further on current battery technology. The range challenge of electric cars would already be solved.

Contrast that with the self-driving car technology developed by Waymo—Google's sister company. One of its promotional videos shows two friends programming the location of a TexMex takeout, by means of voice command. The car rolls out of the driveway and heads into traffic. It deftly navigates its way to the order window, where our friends pick up their meal. Chatting and chewing, they head back to their starting point. Once the car is in the driveway, the person in what we will quaintly refer to as the driver's seat opens the door, unfolds his blind person's white cane, and haltingly sets off to the front door of the house.

The film ends there, but the next scene could well have shown the car heading off to pick up someone else to take him or her to the next destination, obviating the need for a parking space. When the car is not required, it can go recharge and park itself fender-to-fender in a remote or subterranean garage. Just as 80 percent of airplane crashes result from pilot error, the vast majority of car crashes also stem from human error.[7] Early data from several million kilometers of fully autonomous driving appears to confirm the expectation that robot cars can eventually be made very substantially safer than those driven by humans.[8] Safer cars would alleviate the need for the heavy steel reinforcements. Moreover, while heavy steel armor protects those inside the car, those outside benefit from lighter vehicles. Lighter cars use less fuel. All that could lead to a virtuous circle of fewer crashes linked to lower energy usage—and cost.

Back in 1993 Rocky Mountain Institute founder and polymath Amory Lovins argued that it was in principle possible to build a car that weighed 1,000 pounds and got 150 miles to the gallon. It would retain its safety and performance features. But it would require so many changes in the design, manufacturing, and logistical processes in the auto industry that it would require a revolution in the industry. And so, predictably, it did not happen.[9]

Thus the fully robotized car has potentially a very different end point than gradually adding driver-assist features. As with anything in the future, the path is speculative, but a direct comparison shows how profoundly different the two might be (see table).

	Incremental *Driver-assisted individually owned automobile*	*Revolution* *Fully autonomous publicly shared vehicle*
Car production	About the same	Many fewer cars produced
Energy usage	About the same	Up to 90% reduction
Road safety	Some reduction in fatalities	A small fraction of accidents
Range anxiety	Wait for improved batteries	Solved with current batteries
Urban impact	About the same	Up to 30% of urban space freed up for rethinking cities
Cost of ownership	Electric cars more expensive	Largely unknown
Environmental impact	Some improvement	Much shorter product cycle, rapid implementation of innovation
Ownership	Largely individual	Largely collective—pay per use
Social norms	Car remains a visible element of social identity	Hard to predict as consumer social norms coevolve

The path in the center column of the table is the incremental one, and the one on the right the revolutionary one. While the latter holds great potential, it is not painless. It will certainly be highly disruptive to lots of players along the traditional automobile industry's supply chain. The incumbent car industry will doubtless have a strong preference and advocacy for the incremental path, as it will be less disruptive to its way of operating. And as we saw in Chapter 10, companies as institutions are not designed to change very much. However, such is the nature of technological progress. After all, the car industry itself displaced the horse and buggy industry that came before it. The opportunities for climate policy are largely concentrated in the revolution pathway, both in terms of direct benefit through energy saving and material reduction, and by opening new opportunities for social norms to evolve and cities to be rethought.

A Transport Revolution

The effect will go well beyond the car industry. Transport writ large will never look the same, nor will everything linked to it—from hotels and housing to the way we consume wine, news, and entertainment. The social disruption could be substantial. Driving a truck, bus, or taxi is one of a diminishing number of jobs open to many without a college education.[10] Breaking path dependency will extract a toll, which needs to be taken into consideration.

Contrast those disruptions with potentially enormous societal benefits. It begins with countless lives saved from tragedy and commuting time freed up for leisure, sleep, or work. Getting there will not be a straight path. Witness the upheaval about the first casualties from autonomous vehicles being road-tested, in stark contrast with the daily carnage from human drivers. Testing, too, takes wise regulatory oversight and involves hard questions of balance between allowing and encouraging technological progress, all balanced against the costs.

Meanwhile, environmental impacts, both during the transition and in the final state, are uncertain. For one, the material energy intensity of the transport system might well plummet—fewer, lighter, and less-thirsty cars. That is clearly good, and the increased productivity—more economic output with fewer inputs—is surely positive. On the other hand, driving just became a lot "cheaper," in the all-encompassing meaning of the term.[11] That potentially leads to more of it. People could send the car to pick up some milk or a pizza. While this would generally be good for society, as people vote with their feet, or in this case their wheels, it may well be bad for the environment. It all still comes down to the near-permanent tug-of-war between human ingenuity and environmental limits.[12] There are opportunities aplenty, such as the opportunity to fundamentally rethink the way cities function, as parking areas and roads are freed up for redevelopment. But laissez-faire won't do. It may take enlightened—revolution—policy to realize its full potential.

Although self-driving cars are an exciting development in their own right, the concern in this book goes beyond this example. It is to illustrate how policy makers could and should approach this kind of situation where there is an apparent choice between an *incremental* path and a *revolutionary* one. They might be tempted to take a neutral stance, fielding jejune arguments that the government should not pick winners and should allow the market to decide. But this position willfully ignores the reality of path dependencies, network effects, and lock-ins.

Many policy problems have such dependencies. In their presence, applying a laissez-faire approach has the overwhelming likelihood of favoring the incremental pathway, in our example, the driver-assist solution. Free market advocates vehemently reject picking winners, but in such cases, laissez-faire is implicitly picking a winner, by choosing not to intervene. Nonaction is not neutral. To create a true level playing field for these two pathways, policy makers need to develop the skills, the language, and the tools to deal with the complex nature of the system. In some cases this will require consciously severing some path dependencies. Such actions need to fit within existing governance systems and be explainable to various stakeholders. But leaving the market to sort it might well be implicitly picking winners.

Lots of new policy thinking is required, some obvious, some less so. Take insurance. With autonomous cars, there is always the question of who is liable in the case of an accident. The answer is simple: surely the company that makes the product. But if there are far fewer accidents, then also the aggregate cost of insurance will be much lower. In the United States alone, a 90 percent reduction in accidents would save almost $200 billion annually. Insurance, after all, is just a way of spreading costs among those who face a risk that is relatively rare. Today, insurers really cover the cost of the consequences of human error. With autonomous cars, this shifts to insuring a risk of technical failure.[13] While this is a huge shift for the current car insurance industry, which would go from covering billions of people to only a handful of companies, it is a product liability cover the industry already provides in many sectors. So the insurance industry could accomplish this shift without too many difficulties.

But there may well be another less visible path dependence that requires attention. Insurance is only one part of a long list of existing regulatory lock-ins and issues. If the risk is ten times smaller, then shifting the risk from the driver to the manufacturer will be ten times cheaper. But there may be a hiccup. In practice, courts may well treat corporations differently than they treat the sum of the individuals. Penalties and liabilities could be substantially higher, as a heavier guilt is imputed to deeper corporate pockets. Many costs are now not internalized into the cost of driving—but covered elsewhere. Medical insurance covers some risks; the state covers a lot of expenses tied to road accidents. Other costs such as delays from blocked roads are never recovered and are simply borne by traffic participants. All this could change as the deeper pockets of the car manufacturers come into the sights of the litigators. As a result there is a real danger that many of the enormous potential savings from radically higher safety could disappear as other costs are reallocated to the manufacturers. Even the uncertainty about whether this can happen will function like a stuck handbrake on a tractor.

Bryant Walker Smith has chosen a rare academic specialization:[14] he is one of the leading experts on the legal aspects of autonomous cars. As such he can cite a long list of obstacles that need to be removed for the technology to flourish, such as the fact that buried in New York State law is a requirement to have one hand on the steering wheel at all times. There are many other path dependencies to deal with. Unknown to most drivers is that they cruise under the umbrella of the United Nations' 1968 Convention on Road Traffic. In March 2016, sponsored by five European governments, the convention was modified to allow for autonomous vehicles.[15]

Dealing with this path dependency is tricky. It can look like shielding industry from liability, when the real intent is to open a path to a revolution through a direct shift in cost from consumers to companies. Such asymmetries in blame and liability are real in society—breaking them is subtle but essential work.

The main reason cars are heavy is because they are certified to drive at very high speeds and have accident-prone agents at the controls. All cars are designed to go well above usual maximum speed limits.[16] Yet to be (relatively) safe at those high speeds requires a disproportionate amount of heavy steel protection. Autonomous cars no longer need all that weight, if they rarely crash. But there is a big caveat. They cannot mix with the heavy and more erratic human-driven machines, as these would form an enormous danger for the nimble robots. Waiting for the old cars to disappear from the roads would take decades and virtually guarantee a lock-in of heavy autonomous cars. The implication is that a radical, expensive, and difficult step to take would be to reserve separate lanes for cars with and without drivers. Of course, I realize this is excessively difficult—but the real debate should first be whether the danger of lock-in justifies such a radical step in the first place. This is again the kind of nonlinear trade-off that our current policy economics are ill equipped to handle.

Revolutionary Contagion

Taking the revolution path to autonomous vehicles would radically change transport. But our interest is broader. Like energy, transport is deeply interwoven with other systems. How can a transport revolution unlock some of these systems, so as to enable progress on dealing with the climate crisis?

For most complex societal issues, there are typically plenty of network effects, from vastly different network shapes. Transportation is a case in point, so policy makers eyeing benefits from the transition to autonomous cars need to be network savvy.

It starts with the physical road transportation networks themselves. The streets, roads, and highways form a mesh, itself interconnected with other transportation networks such as trains, air transport, and waterways. On the roads there are multiple types of users: trucks, cars, bicycles, e-scooters, pedestrians, and so on. Road users influence each other, most annoyingly when traffic jams occur seemingly out of nowhere. The need for parking is deeply interconnected with cities, but also with the economy through the various subsidies it attracts. Gasoline taxes are one of the cheapest and most effective ways to collect taxes for governments, requiring little overhead and with high collection rates. But shifting to electric transport shifts the tax base and possibly incurs much higher collection costs. Fewer accidents will free up capacity in the medical system. Hotels may get new competition in the form of Autonomous Travel Suites—or rooms on wheels.[17]

You can't drive and drink. You shouldn't drive and talk on a cell phone—or text. Your parked car takes up city space. Human driving impacts the medical system. And the insurance system. Time spent driving could be spent on that book you bought. Car choice can be a big source of social identity.

We could go on, but you get the point: interconnections between systems matter a lot. Devising policy for societal issues such as autonomous cars will require a good look at many of these interconnections—but also at the structure of the various networks, so that we can project how they might evolve under these interventions.

In the table above we contrasted two pathways to autonomous cars. On the one hand, we consider the incremental path where cars would gradually adopt driver-assist features, making driving more comfortable and safer, with an ultimate goal someday of dropping the steering wheel. They would continue to be mostly individually owned, which means mostly parked, as well as drivable by humans, which implies heavy protective steel. On the other hand, we have the revolutionary path where cars are fully autonomous as quickly as possible, essentially on a pay-per-use basis and insulated from heavy behemoths driven by fallible humans. These cars would be lightweight and thus energy efficient, on much shorter life cycles because almost continuously in use and likely much safer than current automobiles,[18] addressing the third cause of injury in the United States, sending almost four million people a year to hospital emergency rooms.[19] Existing road capacity would be enormously increased and inner cities transformed as vast amounts of parking space become available for redevelopment.

Sure, both pathways will deliver more comfortable and safer ways of getting around, but only one has the potential to redefine our cities, help fix the climate, and provide new freedom for how we shape our world. The prize is enormous, in terms of quality of life as well as environmental benefit—but it is attainable only through revolution policies, as it will otherwise be held back by the gravitational pull of path dependencies. Like a rocket reaching escape velocity, you need to overcome the ties that hold back.

Path dependencies on the incremental path might well hinder ever reaching the full potential of the technology.

What measures might have an effect comparable to closing Germany's nuclear fleet, and what norms policy can be considered? Just to be clear, we are not claiming to resolve a complex policy question such as autonomous vehicle policy in the short context of this book. Such problems cannot be "solved," as they are so deeply codependent on systems other than transportation alone. This requires lots of experimentation and adaptive policy making. Also, it would take us to a technical level that is not appropriate for our main aim of providing a relevant illustration for the revolution approach to policy.

Under the revolution pathway cars are shared. Not only driving, but also ownership disappears, vacating a large area of social identity, with an enormous potential for new norms to emerge. Notwithstanding the hype about the sharing economy, there is no doubt that personal ownership of cars is really sticky. And getting autonomous cars to be shared and not individually owned is one of the foundations of our revolution pathway. *New York Magazine* reflects on the loss of driving:

> The experience of driving a car has been the mythopoeic heart of America for half a century. How will its absence be felt? . . . Will we mourn the loss of control? Will it subtly warp our sense of personal freedom—of having our destiny in our hands? Will it diminish our daily proximity to death? Will it scramble our (too often) gendered, racialized notions of who gets to drive which kinds of cars? . . . What will become of the cinematic car chase? What about the hackneyed country song where driving is a metaphor for life? . . . Will we all one day assume the entitled air of the habitually chauffeured?[20]

The same is true today in many emerging economies, with one important difference. With the much higher population densities than fifty years ago, individual car ownership everywhere is leading to traffic coronaries in developing megalopolises, as road capacity falls well short of demand.[21]

There are some early signs that the attractiveness of individual car ownership is already eroding, such as the increase in age at which Americans obtain their driver's licenses.[22] What happens is that young people get their licenses later, and in the meantime get used to taking other forms of transportation. They use buses and bikes more often. By the time they do get a car, they use it less compulsively. It turns out that the age at which you pass the driver's test is a strong indicator for how much you will drive later in life. So one avenue for loosening the ownership norms is nudges for getting a later license. This sounds difficult, but in fact the United States has already made getting a license harder.[23] There is scope to go further. It is well established that using a cell phone while driving—even hands-free—is the equivalent of imbibing a few beers, so it is a ban waiting to happen.[24] In some companies, it is virtually a fireable offense to use any cell phone while driving, for reasons of safety.[25] In short, the arsenal for discouraging early driver's licenses is not exhausted by any means. Making this an explicit policy for breaking a path dependency provides a broader justification.

Another example is UberPop. When Uber tested various new services while it was exploring its innovative business models, one of its most controversial offerings was UberPop. The idea was to allow everyone the opportunity to use their own car to moonlight as a taxi driver. Using the Uber app, you could opt for a professional driver—or a private car. For example, in Amsterdam people really liked the service. It was priced at half the cost of a normal taxi and had no accidents, as well as a colorful cadre of drivers:

> "One was a late 20s vegan chef who was making money on the side. Another was in his 50s and was on disability. He had been home after an injury for five years. 'Who is going to hire a 50-year old with health problems? No one.' If his back felt good, he continued to take rides. When he got tired, he explained, he went home. 'I feel like I'm back in the game,' he said with a huge smile."[26]

Yet the courts ordered the Amsterdam service to shut down the system in 2014. Uber continued for almost a year, picking up the fines, but had to bow out in the end.

UberPop had led to a predictable uproar with taxi drivers, as well as a tidal wave of lawsuits in various countries. All sorts of objections were raised. Pop drivers wouldn't know the way; they lacked training, were unlicensed, unsafe; it would destroy the livelihood of existing taxis, and so on. None of these objections holds up particularly well to close scrutiny, except probably the last one. As in many markets, a set of habits and regulations had been built to protect the incumbents. This may be justified in some cases, as society balances economic efficiency with some level of job stability. But the result was that in most jurisdictions Uber stopped the UberPop service, to concentrate on its other offerings.

The one thing never considered was that allowing UberPop might well have been a norms policy preparing the way for autonomous cars—a norms policy that helped loosen the individual ownership model of cars, as more and more people came to reframe them as a shared resource. This might occur either by picking up passengers themselves, or by more people using the cars of others. Like the German solar feed-in tariffs, it might have been a catalyst in preparing the way for the autonomous car revolution. It's too early to tell, but it may well be that the rapid growth of ride-sharing apps like Uber and Lyft in the United States are preparing the ground for a more rapid shift in social norms than otherwise would occur. When taxis and Ubers become part of your daily routine, a robot car is just the next step.

Awkward Policies

The potential reward for society from autonomous cars is substantial. But it will require a purposeful choice between an incremental and a revolution pathway. Only the latter can deliver systemic change. The goal is two-fold: a revolution in transport and loosening path dependencies in multiple adjacent systems to overcome some of the lock-ins that stop us from dealing with the climate crisis. It will require turning a deaf ear to the incumbent car industry and designing a host of catalytic interventions that will trigger the revolution, some of which we described, many more to be discovered along the way. However, both path dependency and the power of the incumbents give the incremental path much greater odds. The prize goes beyond a transport revolution. It offers the chance to recast social norms, the design of cities, and their energy consumption. But these benefits are accessible only through the revolution path, not the incremental one.

Revolution policies are challenging, and, unlike the incremental scenario, triggering a revolution requires active policy. Like solar feed-in tariffs or closing nuclear plants. Subsequently, those policies can do their systemic change work quietly and out of the limelight. To deal with the climate crisis we'll need some of those revolutions. Being purposeful about harnessing the winds of change requires complexity literacy and a steady hand at the rudder.

Conclusion

Beyond a Blueprint for the Climate Revolution

The essence of tyranny is the denial of complexity.
JACOB BURCKHARDT, HISTORIAN

Precocious? absolutely. By age three Alma Deutscher played the violin and the piano, at six she had composed her first sonata, and her first full opera debuted in Salzburg when she was twelve. Wonder children deserve outstanding tools, and Alma joins the great violinists through being loaned a seventeenth-century Italian instrument—in her case a Bergonzi. Bergonzi, like the Stradivarius, Guarneri, and Amati violins from Cremona, all deliver unrivaled sound and fetch fabled fortunes at auction. Why? One explanation is global cooling. For a century starting around 1580, the world's climate cooled by 2 degrees. This had enormous knock-on effects, among them that trees grew more slowly, growing their rings more tightly— which in turn produced the denser unique wood from which the Cremonese family firms fashioned a legacy of precious instruments.

While the cause of seventeenth-century global cooling is still unclear, its impact was everything but. Some consequences were positive—such as amazing violins or midwifing the emergence of the Enlightenment philosophy—and others were negative, such as widespread famine, military conflict, and huge disruption of economic activity. As hunger struck, peasants became displaced people, feudal landlords lost income, and cities faced with famine rioted. Food was essentially produced and sold local for local, with very limited capacity to manage the larger fluctuations from climate vagaries through trading. In Europe an epochal shift occurred as southern nations such as Italy and Spain fumbled their dominance to the rising powers of the North such as the Netherlands, England, and Sweden. These had developed trading institutions and infrastructure adapted to the age of cooling. Just 2 degrees had changed the social, economic, and natural systems of the continent—and with interconnected effects that escape precise under-standing and modeling, even in hindsight.[1]

Our technical capacity has evolved enormously, but so has our depen-dence on the resilience on systems such as agriculture, energy infrastruc-ture, or the Internet. The experience of the seventeenth century undoubt-edly holds lessons for how today's climate crisis will impact our complex societal systems. The example of the seventeenth-century global cooling suggests that it is plausible that we greatly underestimate both the impact and the scope of both mitigating and adapting to a changing climate. Our growing understanding of complexity could help us cope, but only if we integrate those insights more closely into climate policy.

In Chapter 1 we asked whether we could identify those purposeful in-terventions that change path dependencies. Displacing strong existing path dependencies can unlock the possibility of rapid change. It was emphasized that we would provide neither certainty nor a firm recipe, but as described in the previous chapters, we know enough to describe examples and ap-proaches. Complexity policy offers a different framing, not a menu of so-lutions. This is a necessarily humble approach—but that is the consequence of embracing complex systems with their irreducible uncertainties. Clearly, science will progress and provide better models of complex system transi-tions, but we can already evolve our decision making today. Indeed, the climate crisis makes this imperative.

Fixing the climate crisis requires breaking lock-ins and overcoming path dependencies in multiple human systems simultaneously, including food, energy, and transport, through interventions that change incentives, in-terests, norms, and options. Economic assumptions routinely carry over unquestioned or unarticulated into policy.[2] Indeed, while economics may sometimes have more nuanced views to offer, those often linger at the cutting edge of science and do not influence policy for decades. Hidden assump-tions that sit uncomfortably in complex systems underpinning concepts such as equilibrium, representative agents with average properties, fixed social tastes and norms, pricing as a way to match supply and demand, and externalities all deserve to be made more explicit in policy. Whenever you are confronted with an average statistic, develop the reflex to question what diversity is hiding behind the smoothed-out number. Challenge your-self on whether you have considered path dependencies and network effects. What is the specific network topology that may be at work—or that might be coaxed into existence? Unearthing and articulating those foundational assumptions is a start.

In this Conclusion we describe some of the ingredients that are required for systemic change. We don't pretend it to be an easy how-to guide. The complex nature of the underlying dynamics would make that nonsense. Irreducible uncertainties will continue to loom large. But that doesn't mean we cannot purposefully aim for nonlinear change, albeit with much attention to how to organize this by setting up timely feedback loops, making course corrections, and managing unexpected consequences. For sure it requires an ability to run small experiments and learn from them. But that is not enough. The experiments need a complexity frame within which they are conducted.

Incremental versus Revolution

In a now classic paper in 1959, Charles Lindblom describes the science of "muddling through." He contrasts two views of public policy and notes that "for complex problems, the first of these two approaches is of course impossible."[3] The first view, the impossible one, consists of exhaustive analysis and picking the optimal proven pathway. It works only for simple systems, which is irrelevant for most public policy issues. The second one, the muddling through approach, consists of setting clear goals and then proceeding with targeted analysis, small experiments, and gained experience. Complexity adds a more precise framing of how rapid systemic change can happen.

Every time you consider a proposal for an improvement, consider whether it is an incremental improvement, but one that possibly even deepens the lock-in of the existing basin, or whether it provides a passage to a new "basin." Many efficiency improvements will be of the former kind. By all means make them, but remain conscious of their limited systemic impact.

Some of the debates about climate policy have been more like Lindblom's first option. Cue the assertive statements on whether a fully renewable power grid is possible, or the overly precise assertions on the long-term economic impact of climate action. Those certainties do not exist for a complex system, and voicing them is at best distracting, at worst misleading. Better to muddle on and learn along the way.

And separately make plans for policy revolutions.

Break Path Dependencies

To break path dependencies is to purposefully block some paths. Like a park ranger cutting trees to make undesirable trails inaccessible and nudge hikers onto the official ones, policy makers can discourage certain paths.

This happens—all the time—but the difference here is that it is done with the explicit intention that the other path will trigger a sudden change. Not simply the selection of one path over another, but blocking a path for incremental change, in order to open a path for revolutionary change.

Singapore is close to perfect for an early implementation of autonomous electric cars.[4] It is small, dense, wealthy, and mostly insular. And whereas Norway has pioneered the mass introduction of electric cars, Singapore has been extremely reticent to incentivize them, either directly or through the deployment of a charging infrastructure.[5] If indeed one's goal is for a collectively owned autonomous car system, this reticence may be sensible. It avoids a possible new lock-in of individually owned electric vehicles—picture heavy Teslas everywhere—in favor of leapfrogging to a collective autonomous vehicle system. Clearly discouraging a new path dependency does not yet a revolution make. But such is the nature of unfolding peaceful revolutions.

Look for Catalytic Interventions

In Chapter 2, we described how suppressing for-sale signs had plausibly been a catalyst to avoid racial segregation in a Chicago suburb. And we characterized this kind of policy as the ultimate prize for policy makers: a relatively small and low-cost intervention that has the potential to change the path of an entire system.

The air quality in Beijing provides another example. It is terrible and has been so for a while. Everyone in Beijing knew, but there was no shared data or mechanism to apply pressure on policy makers and industry.[6] In 2008 the U.S. embassy in Beijing installed a fine-particle monitor that measured how much of the finest particles were in the air. These are the nastiest bits, most devastating to the lungs. The measurements were tweeted out every hour. This small intervention has catalyzed enormous change. Catalyzed, not caused. By the beginning of 2013, following protest and denial, and much huffing and puffing, the Chinese government rolled out more than 500 air quality monitors in seventy cities.[7] The single detector on the U.S. embassy roof has had a major impact on health and the environment in China, and arguably also on its industrial structure and even political discourse—a revolution in the making.

There is no formula for identifying such catalytic interventions. However, complexity-literate policy makers have a much bigger chance of finding them.

Catalysts work or can work. Henry VIII's action to divorce the first of his six wives had a catalytic effect in separating England from the Catholic

Church, which in turn had an enormous impact on British and European history, making him an early if unintentional Brexiteer. Rulers have always used symbols and the theatrics of power as catalysts:[8] small interventions with the aim to reinforce power, such as Napoleon raising the imperial crown to his own head to signal the primacy of his power ahead of the divine; or a freshly crowned president Macron choosing to stand in front of an open window, flanked by both the French and EU flags for his official portrait, to signal a culture of openness. There are many such illustrations, but what we are after in this book is purposeful policy interventions for systemic change. They could well be symbols, or technical interventions such as a single fine-particle monitor, or a concise law like the German solar feed-in tariff. Note that this can work in different directions: symbols as catalysts can lead to policy shifts, and sometimes policy shifts can be catalysts for social norm shifts.

It all starts by acknowledging that catalyzing a revolution is a real policy option, for which we have a rough idea of the mechanics for scale-up and then to hunt for possible candidates.

Goal Setting

The climate goals are clear: keep warming to less than 1.5 degrees. But it took enormous global effort to set these. Parents can set bedtime goals. But what if there are no parents, and the kids need to set their own goals?

DARPA's 2003 challenge for driverless cars to complete a grueling 142-mile race ended up creating a new innovation ecosystem.[9] Responding to a cue from Congress that demanded that a third of vehicles on the battlefield be uncrewed, DARPA invented a race and dangled the promise of a million dollars to the first team who completed it. Tony Tether of DARPA thought that perhaps five or ten participants might show up to the start.

The first race took place on March 13 2004, and DARPA had to select 25 participants out of 106 applicants. The response had been overwhelming. Their performance unfortunately was not. One of the first cars made a U-turn after the second curve and headed back to the finish line. Others were stranded on the rugged terrain. The best car ran a full 7 out of 142 miles before catching fire.

But it was only a failure in the narrow perspective of the immediate result. It was a great success in two respects that are essential for catalyzing a revolution: a goal had been set that was truly challenging, yet credible enough to attract participants. And a community of practice had been created. Several of the people participating in the original race ended up joining

Larry Page at Google for the first concerted industrial effort at building a driverless car. As we saw in Chapter 3, a networked system needs a challenging goal to allow self-organization to occur from the bottom up.[10] DARPA's prize provided that.

DARPA subsequently doubled the prize to $2 million and scheduled the next race. Eighteen months later, 5 out of 195 teams completed the course. Several companies went on to design the intended robotized battlefield vehicles for the U.S. Marine Corps. But while the immediate goal had been accomplished, the necessary conditions for a revolution had also been put in place. And that particular revolution has yet to play out.

DARPA was outside of the system and could set goals through a prize, like a parent setting bedtimes. For climate policy, in many cases everyone, including the government, is inside the system, and the kids are making their own rules.

Palm oil is a nearly ubiquitous commodity that finds its way into myriad consumer products, such as soap, shampoo, processed foods, and, increasingly, biofuels. It is derived from the fruit of palm trees. Scaling up to meet the demands of industrial production required dedicated palm tree plantations, mainly in Indonesia and Malaysia. The clearing of vast swathes of virgin forests, often by fire, is creating enormous smog clouds over Asia and is devastating forest ecosystems that are home to species such as the orangutan. In response the Roundtable on Sustainable Palm Oil (RSPO) was set up in 2009, assembling more than 3,000 members along the entire value chain of palm oil production and consumption.[11] It is often celebrated as an example of successful voluntary action by industry, but unfortunately that is misleading.[12] For example, the standards actually allowed for burning virgin forest for new palm oil plantations. In this case the goal setting was from inside the system, and there was no external parent to set a suitably ambitious level. Goal setting from inside the system had delivered the softest of targets.

The solution is acknowledging the shortfall, agreeing on an interim goal, and agreeing to ratchet. As we saw in Chapter 5, this is what led to the breakthrough at COP21 in Paris. This is much better than agreeing to disagree. It is agreeing to agree more in the future, and agreeing how to do that.

The selection of the 1.5 or 2 degree target itself has a catalytic effect. German climate economist Carlo Jaeger and Julia Jaeger—in a rare father-daughter paper—characterize the 2 degree target as a focal point for action, rather than a quantitative climate threshold: "It leads to an emphasis on implementing effective steps toward a near-zero emissions economy, without panicking in the face of a possible temporary overshooting."[13] They argue that a temperature target is like a speed limit in cities. What matters is to have one that reasonably represents our best knowledge. Debating whether 47 kmph would be a better urban speed limit than 50 kmph does not save lives. The question is whether the target is an effective focal point for action. The climate target may well evolve as new scientific insights arise. For example, the late addition of a 1.5 degree target to the Paris agreement was based on a better understanding of the devastations that the additional half a degree rise would likely cause.

Goals are an important focal point to allow complex systems to self-organize around.[14] Set them from the outside if possible, for example with a prize. If that is not available, then ratcheting is a practical alternative. Setting fixed firm goals from inside the system is a recipe for failure. And recognize that they need to be a pragmatic focal point for action, not necessarily a statement of science with the required nuance and caveats.

Institutions as Guardrails

When you break path dependencies and unleash the capacity for a system to change, the direction the change will take is uncertain. When water leaves its liquid state by heating, it can go only one way, which is into vapor. Not so with societal change. The Cuban Revolution turned a utopian and optimistic vision of the future into an impoverished and illiberal present. The Internet revolution promised to dramatically broaden access to information, but has delivered an echo chamber. What started as a juvenile quest to date hot girls threatens to break democracy, as Facebook's impact spins out of control. One obvious justification for retaining control is that you don't know where things might end up. Unlocking lock-ins is not without risk. Sometimes that is why the locks were put in place. Other times they may have grown out of habit.

Starting from scratch calls for big handrails. It took 150 military personnel, each armed with fifty cans of white paint, to make over the venerable Mount Washington Hotel in Bretton Woods, New Hampshire, in early 1944. The grand hotel is an enormous wooden structure built in 1902 to serve as a holiday playground for the wealthy East Coast set. It had become run down, and closed in 1942. The soldiers painted right over brass fittings, Tiffany windows, gilded ceilings, mahogany doors, and even the Lalique swimming pool tiles. Located at the foot of Mount Washington in New Hampshire's White Mountains, it featured glorious views, easy access, and good security, and, unusually at the time, it admitted Jews. It was the perfect location for imagining a set of new institutions to avoid a rerun of the international financial mess that had contributed to launching the

Second World War. Following the white paint tornado, more than 700 delegates from forty-four nations met for three weeks and designed the institutions that would shape the postwar order. From the debates emerged the International Bank for Reconstruction and Development, the International Monetary Fund, and an exchange rate system based on the gold standard. Ignoring Keynes's advocacy for the British pound, it also anointed the U.S. dollar as the world's reserve currency. Policy makers realized that with the amount of path dependencies broken during the war years, it would require a solid set of handrails to coax the international system back to a positive and stable next phase. The Bretton Woods institutions formed an important part of the new handrails to guide the system to a happy landing.

Similarly, when the German government launched its energy transition, it realized that this would be much more than a transformation of the energy system—it would be the start of an industrial and social revolution. One of the critical steps it took was to provide ample funding to bring into existence a new set of institutions, think thanks, and policy groups. Such institutions provide essential constraints and direction for the self-organization of systems.

The Vision Thing

George H. W. Bush famously groused about whether "the vision thing" mattered. For the climate crisis, it does. From a complex systems perspective we can be precise. A vision acts like the earth's magnetic field for the navigation of migrating birds. They may self-organize into flocks, but an external force helps them with overall direction. In human agents, a compelling vision creates more willingness on the part of the agents in the system to engage constructively in the collective effort.

For climate policy it matters for another reason, as it reduces investment risk. A clear and consistent statement of direction by governments signals to investors that their investments will sail with the prevailing wind, as opposed to against it. That translates into all sorts of assistance, and influences consumer and investor preference, regulatory support, and perhaps even subsidies. Between 2010 and 2016 European power companies have had to take an early write-off on almost €75 billion worth of fossil assets:[15] the cost of sailing against the wind and of lending a tin ear to the vision thing.

China and Europe have found their "vision thing" for the climate, and the United States is working on it. In 2017, in a speech to China's National Congress, President Xi announced that "building an ecological civilization is vital to sustain the Chinese nation's development."[16] This vision is entirely consistent with his nation's naked ambition to dominate the new industries of the twenty-first century, such as electric vehicles and renewable technologies. The European Commission, in less evocative language, articulated a "vision for a prosperous, modern, competitive and climate-neutral economy by 2050."[17] Using a term linguistically rooted in the successful resurrection of the American economy from the economic crisis of the 1930s, some in the United States are attempting to bring to life a Green New Deal as the American response to China's Ecological Civilization.

And while Europe could well use some inspiration from its poets and writers for a better tagline, this battle of big ideas has enormous impact. It will help set the direction for the emergent order of society's complex systems to work their magic. Vision words matter.

Norms Policies

As hard as it is, inventing new institutions seems easy compared to establishing new values. When in 1905 the owners of an electric bus company in London committed fraud, it plausibly changed the course of transportation history. Values and norms matter, in particular during periods of radical change. But they can be influenced through norms policies. As we have described previously, complexity science also sheds new light on how norms coevolve with systems and institutions and how network contagion spreads them. This builds on longer-standing insights from the social sciences and provides a more quantitative framework. It is different from mere propaganda. It is about creating a mechanism for norms to evolve and spread.

We've seen how the institution of the corporation developed over many centuries, since it was first created in early Amsterdam to manage the high risk of cargoes perishing in the lucrative trade to the East Indies. The institution changed and adapted to the spirit of the times, culminating in the 1980s with the centrality of shareholder return as the core organizing principle. An alternative, the for-benefit corporation, was created in the United States and has since been adopted in a handful of other countries. It is not (yet) clear whether for-benefit corporations form an effective enough norms policy. But they are an illustration of the way new institutions can be invented to allow a new set of norms to evolve—beyond the path dependencies formed by the current ones.

Greening supply will not be enough. The climate revolution also requires a norms revolution.

Mind the Collateral Damage

Any attempt to tweak path dependencies or catalyze rapid change will have unforeseen consequences. Some of them may come from mistakes or foreseeable events—which ought to have been avoided through more diligent planning. More likely they will be due to the inherent complexity of the system. Complex systems defy precise forecasting because their deep interconnections lead to small disturbances being propagated in random ways. Autonomous vehicles will unleash a cascade of changes in other systems, inevitably challenging to forecast. How will inner cities develop when parking space becomes available for redevelopment? What will the medical system do with the freed-up capacity from fewer traffic victims—or fewer organs available for transplant? How will the beverage industry be affected when drinking and driving become disassociated? What is the impact on the hotel industry when you can sleep in the vehicle that gets you to your meeting in the morning? How do social norms change when the largest consumer purchase is no longer an important part of social identity (for example, soccer moms without white Volvos or midlife men without red Porsches)? Technology tends to exacerbate social inequality; will autonomous vehicles follow suit?

There will also be consequences to any major change. Autonomous vehicles clearly threaten to displace the jobs of millions of professional drivers.[18] But recall that Uber was forecast to put taxis out of business, whereas in fact it has largely ended up creating new demand.[19] Take forecasts of the impact of changes to complex systems with a grain of salt.

A lack of trust deepens lock-ins. In Copenhagen 85 percent of residents believe people can be trusted, whereas in Houston only 36 percent do.[20] This correlates with all sorts of urban features, such as whether people drive alone or use collective means of transportation, whether they live in mixed developments or single-family housing, and their fears about crime and pollution, for example. All these characteristics vastly diverge between the two cities. A reasonable skeptic might observe that correlation is not causation, but complexity scientists have shown that trust is a plausible enabler for social change.[21] It also fits with common sense that people will be more open to changing their ways and trying new things when trust levels are high. System change requires new links to be formed in a network, which continues to evolve its fabric, until a sudden change becomes possible. Trust is part of the micro effects that enable a macro transition. An appropriate level of societal trust is essential for constructive systemic change.

The kind of collateral damage that can occur with systemic change is sometimes foreseeable, sometimes not. There are obvious ethical and ideological reasons to want to mitigate or soften these effects, but there is also a functional reason. Not dealing with these things could hinder the desired change from happening.

To address the climate crisis, we don't say that the solution is eating less meat. We ask what network tweak could lead to the emergent system property of radically lower meat consumption. We don't say that people should drive less. We advocate a measure that leads to less driving as an emergent property of the transport system. Instead of saying we should close all coal plants, we look for an intervention that triggers the closure of coal plants.

Solutions to the climate crisis are not environmental, or even primarily environmental. It is in the interconnection with other systems that the solutions can be found. Inequality is a climate problem. That inequality is intimately connected to dealing with the climate crisis was illustrated by the protests of the Gillets Jaunes throughout France. As discussed in Chapter 11, the continual increase of inequality had sent the less fortunate further away from the expensive city centers, dangling the promise of cheap road access, while the more fortunate commuted to work with subsidized public transportation from their pricy urban lairs. A caricature perhaps, but one with more than a grain of truth. And when the bill for climate action was passed on to those depending on their cars, it further eroded the fraying social compact. Rising inequality is deeply interwoven with climate policy.

Driverless cars can be a climate solution. Legalizing psychedelics following what science tells us can be a climate solution. Studying the network nature of vegetarianism can be a climate solution. Looking for the interconnection between coal plants and the rest of society can be a climate solution.

Epilogue

We can no longer save the world by playing by the rules,
because the rules have to be changed.

GRETA THUNBERG, TEENAGE ACTIVIST

We're procrastinating. Reductionism has been both the elixir of our materialistic success and a cause of the climate crisis. Because we thought we were separate from nature, we looked the other way when it started complaining. And arguably reductionism now stands in the way of resolving the crisis. Climate science has integrated multiple disciplines within the natural sciences, but largely failed to integrate the social sciences and the humanities.[1] We've lost the sense for interconnectedness that we seek out in novels, poems, and music or during hikes in nature. Complexity science is a thoroughly Western way of returning to those roots and provides an integrating fabric. It is a discipline of the mind, not of the heart. Yet it ineluctably engages the heart as it starts to lay bare the fabric of society and nature. Goethe—notable complexity scientist *avant la lettre*—put it as follows: "Nature! We are surrounded by her and locked in her clasp: powerless to leave her, and powerless to come closer to her. Unasked and unwarned she takes us up into the whirl of her dance, and hurries on with us till we are weary and fall from her arms."[2]

To solve the climate crisis many things "ought" to change—our relationship with nature, consumerism, consciousness, economics—and new norms need to be adopted, for example, flying much less, no red meat, and so on. But saying those things doesn't make them so. We also need a theory for getting from A to B, from the current crisis to a better future. It is important that people write what "ought" to be. It will be a source of inspiration to many and even of action for a few. But we also need an idea of how the change can happen, of how change can scale nonlinearly. The intention of this book is to explore such revolutions and how to make them happen in the real world.

We saw in Chapter 4 how the second law of thermodynamics laid bare the generative capacity of nature. The climate crisis calls for getting familiar with and using the generative capacity of social systems, and then using them to create emergent change. Ironically, this is a process very similar to the very nature that is under threat. Only we apply those principles to save it—and ourselves.

Amid the doom and gloom of the failure to address the climate crisis, there is reason for optimism. Revolutions are possible. Sudden change can happen. And in that narrow possibility lies the relevance of individual agency. Action from the top is not the only source of hope, as necessary and powerful as that may be. There are people who have thought this through and whom we can consult. They are just not the usual suspects. They are also complexity scientists, anthropologists, psychologists, and those economists who have reflected on the essence of systems. Their contribution will matter, as it will trigger new thoughts, and, importantly, it will strengthen our innate sense for interconnectedness.[3]

We'll need the tools, discipline, and language of complexity to debate the solutions we envisage. It won't do to wave our arms and say it feels right. We've come too far on our collective intellectual journey of grasping the dynamics of complex systems to settle for less.

While we continue to seek top-down action, we recognize that government is inside the societal system and is enmeshed with it. It is not outside, and as a consequence struggles with agency. And climate policy is not just about reducing emissions. Interventions in other, interconnected systems may well be more effective and boomerang back to the energy system. Revolutionizing our consumer norms is a foundational component. Complexity literacy and network literacy are important to give us a shared language for interconnected systems. It is essential to go beyond spouting opinions and enable debate about the best way to trigger different emergent collective behavior.

There are plenty of improvements we can make incrementally, and we should do them all. They include more recycling, turning off the lights more often, turning off the tap when brushing your teeth, shunning meat, doing corporate sustainability, flying less, furthering energy efficiency. They will all help, but they will not be enough. Or they might lead to an early lock-in that cuts off the path to the required state. One critical way they will help is by tweaking social norms, so people are more receptive to revolutionary change. But we need to be clearheaded about which policies lead to incremental change and which have the potential for revolutionary systemic change. Do both, but don't confuse one for another.

Be wary of policy makers putting too precise a number on long-term economic impacts for climate policy. Their models are dominated by discount rates that define how we value the future. Philosopher Tyler Cowen of George Mason University argues that you cannot discount human lives the way you discount investments in a factory. He provides a stark image by noting that if you take a low 5 percent discount rate, this implies that the value of a human life today is worth 132 lives a century from now.[4] Not right. With that logic, no wonder action on climate change is not felt as urgent. Economists have debated this,[5] but Cowen's point shows that this is better left to other social scientists than to economists.

Panaceas are anathema to complex systems. Complex systems necessarily imply adaptive policy, course correcting as new insights are obtained. But it is more than simply muddling on. Muddling on depends on observable outcomes so as to tweak the approach. It assumes mostly linear causality between the measure and the output. In complex systems, the consequence of actions cannot be observed empirically, at least in the short term. It also requires a model of the internal dynamics. Complexity policy is also adaptive policy, but one founded in our growing understanding of the internal mechanism of emergent behavior, of network effects, of path dependencies. It takes those into account for a much richer theory of change than the empiricists can muster. Complexity modeling helps too, at a minimum to better grasp the dynamics, occasionally to provide concrete guidance.[6]

The bottom-up measures will end up making the inescapable top-down measures easier to take. In a complex systems view, the government is inside the system, not outside. It coevolves and is intertwined with it.

All this requires policy makers to know about complex systems. That is not too hard, and it will assist them to reconnect to the innate capacity that humans have to conceive of an interconnected world. And it will give them a language to reason about it. A climate revolution is still possible—only just—but it requires bringing substantially different skills to the table and adopting a fresh language.

NOTES

Chapter 1. Time's Up

1. "Reaching and sustaining net-zero global anthropogenic CO_2 emissions and declining net non-CO_2 radiative forcing would halt anthropogenic global warming on multi-decadal timescales (*high confidence*)" IPCC 2018.
2. We will not review the state of climate science here. Suffice it to say that there is a long list of increasingly exasperated calls for serious climate action. For a comprehensive, consensus-science take, see the succession of formal assessment reports by the Intergovernmental Panel on Climate Change (IPCC 2013, 2007, 2001, 1995, 1992) as well as those by national academies around the world. For accessible versions, see, in chronological order, among many others, McKibben (1989), Kolbert (2007), Lynas (2008), Schneider (2009),M. Mann (2012), McKibben (2013), Kolbert (2014), and Goodell (2017). Perhaps one of the more telltale signs of this exasperation is the increasing attention paid to solar geoengineering (Goodell 2010; D. Keith 2013; D. W. Keith 2000; Morton 2015; Wagner and Weitzman 2015).
3. New York Times Editorial Board 2018.
4. C. J. Smith et al. 2019 confirm that "if carbon-intensive infrastructure is phased out at the end of its design lifetime from the end of 2018, there is a 64% chance that peak global mean temperature rise remains below 1.5 °C."
5. Thompson 2018.
6. Stoknes 2014.
7. Foust and Murphy 2009.
8. Complexity science is not the only new lens that can usefully be applied to the climate crisis. Besides using the natural sciences to understand the evolution of the climate itself, only one narrow branch of the social sciences has been applied consistently, namely economics. More varied voices from the social sciences (e.g., sociology, psychology, anthropology, political science) are needed to address how the problem is accepted by the public and how governments

and the public will respond to the proposed solutions (Stoknes 2014). As Cairney and Geyer (2017) put it: "It is therefore difficult to establish the sense that complexity is a new way of thinking, rather than simply the right way to think. Instead, complexity theory may help us bring together many strands of the political science literature into one framework."

9. This book is not intended to be a critique of economics; rather, it discusses how economics in practice impacts climate policy, where the neoclassical principles mostly inform the political narrative. There are aspects of complexity in many parts of economics. This certainly is the case with the classical early economists, Schumpeter, and also for example with evolutionary economics (e.g., Nelson and Winter 2002). And while all that is useful, it is not realistic to expect those traditions to materially inform policy within the time remaining. Balint et al. (2017) provide an overview of complexity and the economics of climate change. They identify four areas of the literature where complex system models have already produced valuable insights: "(1) coalition formation and climate negotiations; (2) macroeconomic impacts of climate-related events; (3) energy markets; and (4) diffusion of climate-friendly technologies." At the same time, their paper amply demonstrates how much is left to do to fully integrate complexity into climate economics.

10. The Yale Environmental Perfomance Index (Wendling et al. 2018) lists Cuba as 55 out of 180 countries, ahead of much richer South Korea. Forests, marine areas, and biodiversity are all better preserved there than in many other countries. Part necessity, part choice, the results are notable.

11. https://thewest.com.au/news/wa/woolworths-staff-subjected-to-abuse-across -wa-as-plastic-bag-rage-hits-ng-b88881211z.

12. Rosenthal 2008.

13. Notwithstanding early pessimistic reports of widespread resistance, it appears that consumption of plastic bags has fallen rapidly in Australia. This reinforces the point that such policies are hard to use with any precision.

14. Jackson 2016.

15. Shortly after the 2009 financial crisis, Carmen Reinhard and Kenneth Rogoff published an extensive review of eight centuries of previous crises and our collective inability to learn from them. Economic literature describes the dynamics that lead to collapse, such as an excess of debt, but the macroeconomic models do not reproduce those nonlinear features (Reinhart and Rogoff 2009).

16. Colander and Kupers 2014; Arthur 2014.

17. Rotman 2019; Mazzucato 2014, 2018; Wagner and Weitzman 2015; Stern 2016; Farmer et al. 2019; Heal 2017.

18. Perez 2010.

19. Energy transitions on average have historically taken a long time, from twenty to seventy years (Gross et al. 2018). This, however, hides the very considerable difference in speed of transition. Past global energy transitions have taken a long time—so we might conclude that the challenge we face is historically unprecedented and will require historically unprecedented action. This is partly right, but we tend to think of energy systems as being far more static and stable than they truly are. There is no reason to think that energy is immune from the disruptive

forces that have transformed so many industries. History shows us that when change comes, it can be swift and dramatic (Tsafos 2018).

20. See Yellowstone Park (2011) and BBC (2014) for the commonly told story, but Marshall, Hobbs, and Cooper (2013), among others, voice significant criticism about it being an oversimplification. Recent research rebuts some of the criticism (Boyce 2018). See Middleton (2014) for a brief synopsis, grounded in Marshall et al.'s work. The "butterfly effect," too, has garnered more popular and literary than physical support, as the many self-help books and memoirs featuring the words in their title demonstrate.

21. Scheffer 2009.

22. Many accounts have been written about the outbreak of World War I and what followed. See, e.g., Tuchman (1986) for an authoritative guide.

23. *The Economist* focused in particular on European countries as being too close to allow for warfare to tear them apart (1913). The Paris Peace Pact of 1928 similarly did not end wars, though it might have had a much more significant role in maintaining peace than commonly assumed (Hathaway and Shapiro 2017).

24. Chapter 6 of *The Prince*, published in full, posthumously (Machiavelli 1532).

25. There are many useful introductions to complexity and its impact on various disciplines. Waldrop (1992) provides an interesting historic perspective of the start of the discipline. Gell-Mann (1995) provides a broad conceptual introduction. Prigogine and Stengers (1984) explore the implications of complexity and biology and even date the birth of the science of complexity to 1811 when Fourier described the propagation of heat in solids; Beinhocker (2006) draws out the consequences for economics; and Colander and Kupers (2014) do the same for public policy.

26. Colander and Kupers 2014.

27. See Schumpeter (1939) and Schelling (2006). For Acemoglu, see in particular analyses of induced technological change (e.g., Acemoglu, Akcigit, et al. 2016; Acemoglu, Aghio, et al. 2012).

28. Beinhocker's (2006) *The Origin of Wealth* is an example of a potent critique of economics as a discipline, while it often stops short of providing "solutions.'" See Colander, Holt, and Rosser (2004) for more conciliatory perspectives, especially interviews therein with Ken Arrow and Paul Samuelson, chapters 10 and 11, respectively.

29. A common critique of complexity science is that it delivered a couple of big ideas in the 1980s, but what were the achievements of the past two decades? In fact, considerable progress has been made in making more precise and understanding more deeply those original ideas. That is no small accomplishment, as it lays the basis for an experimentally testable science. It is hoped that a platform has been created for new progress in the future. See Thurner, Hanel, and Klimek (2018, p. 398) for a discussion of this. In any case, the consequences from those big insights from the 1980s have not been fully absorbed into policy in general, and climate policy in particular. That is the core of this book, as the more detailed mathematics are out of its scope.

30. Ramalingam 2013.

31. In 2015 a report was produced by the author in collaboration with scientists from the Netherlands Scientific Council for Government Policy (WRR). The Netherlands lagged far behind other European countries in its energy transition, and the study suggested using a complexity lens on the issue. Pointing to the particularly strong carbon intensity of the economy and the close coupling of the economic sector, it identifies a number of systemic interventions (Kupers, Faber, and Idenburg 2015).

Chapter 2. Getting Unstuck

1. Hamer 2017.
2. The Economist 2007.
3. This in no way implies the reverse, namely, that all segregation is due to systemic emergent effects. Under the Jim Crow system and the shadows it continues to project forward, racism existed and exists. Schelling merely shows that segregation is not necessarily due to strong racial preference, but can come about from systemic effects.
4. See Schelling (1969, 1971) for the original work and technical exploration. Schelling (2006) puts it into broader context.
5. Almost all complex systems involve path dependent processes. In those cases, the law of large numbers does not, in general, hold. The intuition that more samples will increase the accuracy of the average breaks down (Thurner, Hanel, and Klimek 2018).
6. McKenzie and Ruby (2002) provide a detailed account—and important reconsideration—of the "Oak Park strategy."
7. See ibid.
8. Tipping points might be a much-maligned, oft-overused concept, but Gladwell 2000 nonetheless provides an important—and readable—introduction. For more scholarly work directly related to economic policy making, see, e.g., Schelling 2006.
9. Epstein (2008) is a delightful paper listing fifteen uses of models, other than for prediction.
10. "Representative agent" is the term economists use to describe a typical consumer whose attributes are a reasonable representation of the overall population. This approach has the advantage of making economic modeling work possible in the first place. The disadvantage is that it ignores differences that may actually be relevant to collective behavior.
11. This is the case for macroeconomic modeling, although recent steps have been taken to introduce some diversity, such as in multi-agent Dynamic Stochastic General Equilibrium (DSGE) models. In microeconomics, in part through the influence of psychologists such as Daniel Kahneman and Amos Tversky, there is much more attention to the impact of diverse agent behavior.
12. Daniel Levitin makes the quip in *A Field Guide to Lies and Statistics* (2017).
13. Gladwell 2000.
14. Kirman 1992.

15. For a good comparative history of the Willow Run and comparable efforts in Germany and the UK, see Zeitlin (1995).
16. Mishina 1999.
17. See Sivaram (2017) for the specific data, as well as Sivaram (2018) for a broader exploration
18. One of the notable competitors to silicon PV is perovskite, but there are substantial challenges to upscale manufacturing to catch up with silicon, notwithstanding higher energy efficiency (Rong et al. 2018).
19. The National Renewable Energy Laboratory (NREL) maintains a chart of the conversion efficiency of various PV technologies, which shows linear progress since 1976 of just under 1 percent a year. https://www.nrel.gov/pv/cell-efficiency .html, accessed June 28, 2019.
20. Biello 2011.
21. Arthur (1989) offered a first comprehensive theoretical explanation, David (1985) an economic discussion of the famous QWERTY keyboard path dependence, and Page (2006) for one focused on the political implications. Arthur (1994) and Unruh (2000) provide more comprehensive early surveys, the latter focused on carbon. Fouquet (2016) surveys the literature focused on energy systems. Aghion et al. (2014) offer some structured thinking on path dependence and climate change economics.
22. See, e.g., Acemoglu, Akcigit et al. (2016) and Acemoglu, Aghion et al. (2012). Weitzman (1998) provides a theoretical justification for "recombinant" technological change. See also, e.g., Beinhocker's story about the recombinatory nature of the design of the bicycle (2006). Arthur (2009) similarly dissects the mechanisms of technological innovation.
23. The 10 percent drop in U.S. energy-related CO_2 emissions happened between 2007 and 2013. The downward trend has continued since, though less steeply (Feng et al. 2015; Mohlin et al. 2018). Earlier work suggested that most of the drop was due to a switch from coal to natural gas (Feng et al. 2015; Kotchen and Mansur 2016), and newspaper reports at the time clearly supported the view that many observers—perhaps most—ascribed the drop in emissions to the switch from coal to natural gas, with the recession also playing a clear role. Later work has corrected that view and has ascribed as much of the drop in emissions to the rapid growth of renewables (Feng et al. 2016; Mohlin et al. 2018). Note footnote 1 in Mohlin et al. (2018). Also see page 466 of the 2017 *Economic Report of the President*, which draws a similar conclusion (Council of Economic Advisers 2017).
24. See a number of good descriptions of the history of government interventions leading to the eventual shale gas revolution decades later (e.g., Mazzucato 2014; Porter 2015).
25. The author was the head of global LNG for Royal Dutch Shell in 2005 and closely involved with this effort.
26. EPA tackled mercury emissions directly in its Mercury and Air Toxic Standards (MATS), together with a slew of other rules. Pratson, Haerer, and Patiño-Echeverri (2013) discuss the role of MATS and other rules in the coal to natural gas switch. Meng (2016) considers some implications of coal lock-in for natural gas. Also see Fouquet's survey (2016).

Chapter 3. Network Literacy

This epigraph is widely attributed to Eisenhower, but as is often the case with well-worn quotes, it is devilishly difficult to identify the actual source. In any case, the quote itself frames the questions here nicely.

1. While small effects can have system-wide repercussions, this is by no means always the case or straightforward. Watts (2002) p. 5771 shows that in a strongly stylized network, most small changes simply fade away. He acknowledges that this cannot be generalized to all networks: "It is hoped that the introduction of this simple framework will stimulate theoretical and empirical efforts to analyze more realistic network models (incorporating social structure, for example) . . . " The point in this book is that it is possible and that deeper understanding of network structure is helpful, not that it is easy.

2. Walter F. Buckley (1922–2006), an American sociologist, was likely the first to coin the term "complex adaptive system," in a 1968 publication. It is often thought to stem from the early work at the Santa Fe Institute, but that is not accurate. Buckley's key insight is that the dominant framing of social systems as closed equilibrium systems is both at the core of economics and dominating in sociology and psychology. He states, "Society, or the sociocultural system, is not, then, principally an equilibrium system or a homeostatic system, but what we shall simply refer to as a complex adaptive system" (Buckley 1968 p. 490.).

3. Koniaris, Anagnostopoulos, and Vassiliou (2014, 2018).

4. Koniaris, Anagnostopoulos, and Vassiliou (2014) actually coded multiple types of relationships between laws, not simply referral. While that is scientifically important, it does not matter to the more generic discussion we have in this book.

5. Fowler et al. 2007.

6. Barabási and Pósfai 2016.

7. Ibid

8. D. Watts and Strogatz 1998.

9. Resilience is the capacity of business, economic, and social structures to survive, adapt, and grow in the face of change and uncertainty related to disturbances, whether they be caused by resource stresses, societal stresses, or acute events. Robustness is a more restrictive property that excludes adaptation and growth (Kupers 2014).

10. Pimm et al. 2014.

11. The precise analysis is actually more nuanced. Although food webs are generally not small-world, scale-free networks, food web topology is consistent with patterns found within those classes of networks (Dunne, Williams, and Martinez 2002).

12. By "analogy" we mean more than an image or a representation. The simulation of social processes through agent-based models demonstrates how the same dynamic occurs in social systems as in physical ones. It is an analogy and not anything like an actual scientific law, as social systems have large numbers of parameters that can influence the outcome. Social systems are also fundamentally open systems, which are influenced through their interconnections with

other systems. In a physical phenomenon where there is a phase transition, the number of parameters is small and the system well isolated from its surroundings. As a consequence it is relatively straightforward to describe phase transitions in relatively simple systems, but it is not realistic to do so for very complex systems such as a coral reef or political preference. Nevertheless, understanding the principle at work offers a more precise lens on sudden change phenomena.

13. Martínez 2005.
14. Arthur 2009.
15. Riet and Lejour 2014
16. https://www.rbf.org/mission-aligned-investing/divestment accessed on October 18 2019.
17. The central idea of network theory that scale-free networks are a representation of most complex systems in the real world is not uncontested. Scale-free networks are characterized by power laws. Klarreich (2018) reports in Quanta Magazine that in a study of nearly 1,000 networks, 67 percent could not be accurately characterized by a pure power law. This has led to a fierce debate in the network science community. It likely portends a more nuanced theory of networks, with scale-free networks not being the universal law that it was initially believed it was. However, the considerations for policy hold up in first-order approximation, awaiting further refinement of network theory. The original paper by Broido and Clauset (2018) states that "furthermore, evidence of scale-free structure is not uniformly distributed across sources: social networks are at best weakly scale free, while a handful of technological and biological networks can be called strongly scale free. These results undermine the universality of scale-free networks and reveal that real-world networks exhibit a rich structural diversity that will likely require new ideas and mechanisms to explain".
18. In Ansar, Caldecott, and Tilbury (2013) the authors didn't apply network theory directly, and limited their study to relaxing the perfect rationality assumption of most market models. In their model they assumed agents with bounded rationality in an equilibrium model. They showed that under those assumptions divestment campaigns would not have a permanent effect. Adding a component of social norms that spread among investors to their model does provide for a mechanism for a permanent effect—but only if the norms spread quickly enough. In this way they were able to include a crude model of network effects. They applied this approach to nine previous divestment campaigns.
19. According to Ansar, Caldecott, and Tilbury (2013), total holdings in fossil fuel companies by pension funds and university endowments are $240 billion–$600 billion, while total market capitalization of just the publicly listed oil and gas companies is $4 trillion. This does even not include coal, power, etc. Since in past divestment campaigns only a small proportion of the stocks were actually withdrawn, it becomes clear that actions of university endowments and pension funds require a further network effect to have a meaningful system-wide impact.

Chapter 4. Non-equilibrium Is the Source of Order

Waldrop 1992, p. 147.

1. Pangallo, Heinrich, and Doyne Farmer 2019.
2. Prigogine and Stengers 1984, 287
3. In the present view of physics, it is impossible to travel back in time. An infinite entropy barrier separates possible initial conditions from prohibited ones. Because this barrier is infinite we will never be able to overcome it (ibid., 278).
4. Gardner 2005.
5. Walras 2013.
6. Beinhocker 2006
7. Blanchard 2000.
8. "On November 6, 2014, Kenneth Arrow gave an evening keynote talk at the Santa Fe Institute's Annual Trustees Symposium. The meeting's topic that year was Complexity Economics and Ken spoke about general equilibrium as a backdrop for the talks that would follow. His lecture was detailed, precise, and lengthy, and he spent much of it showing point-by-point how general equilibrium did not match with real economies. Ken was shining a keen light on the creation he was best known for: his work in the 1950s that established equilibrium as the basis both of economic theory and of our view of the economy. And he was showing a mismatch between theory and reality. I was struck by the sheer honesty of this. But it was more than just honesty. It was as if Ken, now 93, was looking back on his earlier ideas and was coming from a different way of thinking about them. He still regarded general equilibrium as important and elegant, but he also saw it as an ideal that might live in some Platonic world but not perfectly apply to this one." (Arthur 2019, p. 29)
9. For an illustration of this type of approach, see the description of Krugman's rant in Colander and Kupers (2014, pp. 157–61), as an example of the reaction to complexity.
10. Stern 2016.
11. Shoup 2005.
12. Jacobs 2016.
13. Technically, it's the Law of Compensated Demand: price goes up, quantity demanded goes down—compensating individuals for their loss in the wealth a price increase engenders. The technical term for this type of demand is "Hicksian" demand, distinguished from "Walrasian," uncompensated demand. See any advanced economics textbook.
14. Olson 1976.
15. Edlin and Karaca-Mandic 2006.
16. This anecdote was related to the author while at COP21 by someone in attendance. Although it is consistent with Musk's idiosyncratic approach, it has not been possible to confirm the speech.
17. Wagner et al. 2015.
18. The term was introduced in Colander and Kupers (2014).
19. Jaeger et al. 2011.

20. We note the distinction between the "Pigouvian" approach, which directly affects pollution prices through taxes or subsidies, and the "Coasian" approach, which directly affects pollution quantities, while allowing for these quantities to be traded (Sterner et al. 2019). Although much is made of the difference in political perception, they both have the same purpose.
21. Ball 2018.

Chapter 5. Top-Down and Bottom-Up Order

Ostrom, Elinor. 1990. *Governing the Commons: The Evolution of Institutions for Collective Action*. Cambridge: Cambridge University Press. Reproduced with permission of Cambridge University Press through PLSclear.

1. "Shared Space am Grazer Sonnenfelsplatz eröffnet." https://steiermark.orf.at/m/news/stories/2505084/, accessed November 20, 2018.
2. Cohn et al. 2019.
3. Colander and Kupers 2014.
4. Climate activist Naomi Klein points out that this debate actually is not so much about the state versus the market as about maintaining the status quo: "When it comes to the real-world consequences of those scientific findings, specifically the kind of deep changes required not just to our energy consumption but to the underlying logic of our economic system, [the Heartland Institute] may be in considerably less denial than a lot of professional environmentalists, the ones who paint a picture of global warming Armageddon, then assure us that we can avert catastrophe by buying 'green' products and creating clever markets in pollution" (Klein 2011).
5. UNESCO description made available under license CC-BY-SA IGO 3.0.
6. Lansing et al. 2017.
7. Also see, for example, Simon (1947); Lindblom (1968); Sabel and Zeitlin (2012).
8. Dusza 1989.
9. Colander et al. 2010.
10. Gao, Zhang, and Zhou 2019.
11. Brian Walker tells this story in an educational video recorded by the Stockholm Resilience Centre (Walker 2009).
12. Coates 2015.
13. With a 95 percent reduction of emissions by 2050 over 2009 levels.
14. Klimaat Akkoord 2018.
15. The Economist 2018a.
16. As much as any violence is condemnable, Pinker describes how various types of violence in society have fallen by two orders of magnitude over the past centuries (Pinker 2012).
17. Prigogine and Stengers 1984, p. 313.
18. The tragedy of the commons is a term coined by economist Garrett Hardin. He describes the consequences of a community sharing access to a common piece of land where they have the right to let their cows graze. "Therein is the

tragedy. Each man is locked into a system that compels him to increase his herd without limit—in a world that is limited. Ruin is the destination toward which all men rush, each pursuing his own best interest in a society that believes in the freedom of the commons. Freedom in a commons brings ruin to all" (Hardin 1968). Ostrom's work is in many ways a reframing of Hardin's as a special case of failure of collective action, rather than a general theory of collective action.

19. Ostrom 2009.

Part II. Revolution Policies

1. See Groot (2018) for an example of a detailed road map for the electrification of Dutch industry.
2. Energy Transitions Commission 2018.

Chapter 6. Kicking the Coal Habit

"C'est compliqué de se projeter dans l'avenir quand ce qui vous a constitué s'est totalement arêté." From interview on RTL, https://www.rtl.fr/actu/debats-societe /jean-francois-caron-maire-d-une-ville-pilote-du-developpement-durable -7789403696, accessed November 27, 2018.

1. A complete database of every coal plant is available at http://coalswarm.org/.
2. Sengupta 2018.
3. See, for example, https://www.iea.org/geco/emissions/.
4. Description from https://whc.unesco.org/en/list/1360 under license CC-BY-SA IGO 3.0.
5. March 13th, 2016, Ohio CNN TV Democratic Presidential Town Hall.
6. French coal consumption has slowly dropped by over half between 1990 (when it used 20 million tons) and 2017. CEIC, https://www.ceicdata.com/en/indicator /france/coal-consumption, accessed May 24, 2019.
7. Jun and Zadek 2019.
8. See https://www.banktrack.org/page/list_of_banks_that_ended_direct_finance _for_new_coal_minesplants, accessed December 3, 2018.
9. Ernst and Young provides an overview of European plant impairments: "Is Any End in Sight for Power and Utilities Asset Impairments in Europe[?]," https://www .ey.com/en_gl/power-utilities/is-any-end-in-sight-for-power-and-utilities-asset -impairments-in-europe, accessed January 21, 2019.
10. https://poweringpastcoal.org/.
11. Bundesministerium für Wirtschaft und Energie (BMWi) 2019.

Chapter 7. Wrong for the Right Reason?

Garfield 2018.

1. Energiewende has led to many (English) analyses, together with hundreds of German-language articles. Good starting points, in English, are Beveridge and Kern (2013) and Kreuz and Müsgens (2017).

2. Wagner et al. (2015) provides some summary figures and outlines of the argument. Sivaram (2018, 2017) has the more detailed story of the Energiewende's impact on solar PV.

3. Agora-Energiewende notes that the price of consumer electricity increased by 68 percent between 2003 and 2016, while the underlying gas price increased by only 36 percent. A breakdown shows that the main factor is the contribution to the Energiewende (Graichen and Lenck 2017).

4. See, for example, Agentur für Erneuerbare Energien (2017) for the latest iteration of its annual survey. Remarkably, two-thirds of those surveyed support construction in their immediate surroundings, negating the not-in-my-backyard effect. The institution itself, of course, is a pro-renewables organization, founded in 2005 during the early stages of the Energiewende.

5. Chapter 4 explored the crucial difference between "subsidies" that fund learning curves and those that simply underpin structurally loss-making operations.

6. Baake 2013.

7. German power prices were close to or slightly above comparable prices in other European countries (Grave et al. 2015).

8. There is a cumulative value of EU interventions for coal of $€_{2012}$ 380 billion in the period 1970–2007, the majority of which was in Germany. For renewable energy for the period 1990–2007, cumulative interventions totaled about $€_{2012}$ 70 billion–150 billion, again most of them in Germany. Around 40 percent of that intervention in the EU went to biomass, 25 percent each to wind and hydro, and 10 percent to solar. The reported cumulative Research, Development and Demonstration (RD&D) expenditure by EU member states in the period 1974–2007 was $€_{2012}$ 108 billion. Historically, the nuclear sector has received around 78 percent of the funding, of which the majority was spent on nuclear fission. The remaining RD&D expenditures were divided about equally between renewable energy (12 percent) and fossil fuels (10 percent) (Alberici et al. 2014).

9. The Economist 2013.

10. Der Spiegel 2011.

11. See https://www.wunderlandkalkar.eu/en.

12. Ibid.

13. As with much of the nuclear industry, costs are hard to pin down, and large cost overruns are a regular occurrence. Globally only sixteen plants have ever been completely decommissioned, most of them in the United States, and there is not a solid grip on the actual cost of dismantling. The private operators may not have enough capital in reserve—or may be financially weakened, as in the case of the German utilities. In any case, it is clear that the risk rests with the taxpayer (OECD NEA 2016).

14. See IEA statistics or, for example, Evans (2015).

15. A government report showing a German path to climate neutrality in 2050 (Umweltbundesamt 2014) preceded the political announcement (Deutsche Welle 2019).

Chapter 8. Norms Policy

1. Christakis 2007.
2. An early study of the evolution of social norms using an evolutionary approach can be found in Axelrod and Hamilton (1981). Florini (1996) applies the approach to the evolution of international norms.
3. The Economist 2017a.
4. See Slutkin (2015). This essay provides a discussion and an overview of the relevant literature on the epidemiological aspects of violence.
5. Quoted in Radnofsky (2016).
6. The obvious difference is that the flu is stopped by the immune system and proper hygiene, and violence can to some extent be stopped by individually willed decisions. But that is a difference of degree and not principle.
7. Young 2017.
8. See Graziano and Gillingham (2014).
9. https://www.volunteeringinamerica.gov/data.cfm.
10. https://www.volunteeringinamerica.gov/rankings/Mid-size-Cities/Volunteer-Rates/2014.
11. See Müller and Rode (2013).
12. For an example of an analysis that does not take into account the spatial distribution of solar installations (i.e., the network structure), but reasons strictly based on averages, see Schaffer and Brun (2015).
13. Social norms are excluded from consideration in standard macroeconomic analysis. Nobel economist George Akerlof pointed to this major deficiency in his address upon becoming president of the American Economic Association. In the address he assumes that norms are exogenous and goes on to describe how that assumption contradicts several basic assumptions of economics (Akerlof 2007). David Wilson, who quotes Akerlof, states about the exclusion of social norms and the assumption of perfect rationality that "these absurd assumptions were driven not by ideological bias but by the tyranny of mathematical tractability. The theory couldn't be pushed in the direction of common sense because it would become impossible to grind through the equations. Yet, the theory commanded enormous prestige because it represented the ideal of a science of economics comparable to the science of classical physics. Abandoning the ideal would knock economics off its pedestal and back among the ranks of the other human-oriented sciences such as sociology, psychology, political science, and anthropology" (Wilson 2016).
14. There are multiple such statements in magazines, such as this one in *The Atlantic* (Hamblin 2017).
15. Global warming potential over twenty years according to the IPCC and others.
16. The Economist 2018c.
17. Conveniently, Wikipedia provides a sourced overview at https://en.wikipedia.org/wiki/Vegetarianism_by_country, accessed December 14, 2018.
18. Hsu 2019.

Chapter 9. Kicking the Consumption Habit

This is the closing statement from an interview at 10 Downing Street with Ronald Butt of the *Sunday Times* on May 3, 1981 (Butt 1981). Reproduced with permission of the estate of Lady Thatcher from https://www.margaretthatcher.org/document /104475.

1. The planetary boundaries framework is a representation of the many limits to growth, in addition to the absorption capacity of the atmosphere (Rockström et al. 2009).
2. The IPCC sixth assessment report will start to address demand-side solutions in more detail, which had not happened in earlier reports (Creutzig et al. 2018).
3. Rand, Greene and Nowak 2012.
4. This quote is from June 1930 when Keynes held a lecture in Madrid on 'Economic Possibilities for our Grandchildren' (Keynes 2010).
5. Radden Keefe 2017.
6. The source is Centers for Disease Control and Prevention data, https://www .cdc.gov/nchs/products/databriefs/db329.htm, accessed December 15, 2018.
7. While $1.3 trillion represents only 1.6 percent of the world economy, it is highly concentrated in those areas where the need requires most stimulus. Cement requires little marketing, high-end whisky a lot (PQ Media 2018).
8. UNEP's *Inclusive Wealth Report 2018* shows that 44 out of the 140 countries have suffered a decline in inclusive wealth per capita since 1992, even though GDP per capita increased in all but a handful of them (UN Environment 2018).
9. Emanuel Faber at B-lab Conference in Amsterdam, November 28, 2018.
10. See http://www.societabenefit.net.
11. This history is extensively related in Colander and Kupers (2014). We reproduce only the high-level argument in this context.
12. Stout 2012.
13. Bcorps is a voluntary global certification scheme that effectively does the triage described above. The for-benefit corporation is a legal entity structure that integrates social and environmental goals. In complexity terms, the latter changes the ecostructure, and the former adds a label to the existing ecostructure. Both are useful and related, but each performs a different systemic function.
14. See registry at http://www.societabenefit.net/elenco-delle-societa-benefit/, accessed March 5, 2019.
15. M. Davis 1996.
16. The category of psychedelics (from the Greek for "soul-manifesting") includes a range of molecules that are found in plants all over the world. LSD is a synthetic form. These substances share a couple of characteristics in the sense that their use is nonaddictive and has very limited adverse health effects, they have a very strong effect altering human consciousness, and they have been part of the spiritual tradition of most cultures. One of the unsolved ethnobiological questions is why plants evolved the capacity to have this effect on mammals in the first place. Studies have shown that psychedelics may have a strong effect on treating addictions to substances such as alcohol and opiates. Most of these

psychedelic substances are currently illegal under UN conventions, with some exceptions for research or religious reasons.

17. Peter Schwartz was one of the leaders of the Shell group that developed global scenarios. One of the interesting lesser-known characteristics of those managers is that almost without exception they had some sort of intense spiritual practice that helped inform their work (Wilkinson and Kupers 2014).

18. Pollan 2018.

19. There is no known overdose, although there are some rare cases of mental health issues that appear to be related to existing conditions.

20. https://www.theguardian.com/politics/2009/oct/29/nutt-drugs-policy-reform-call, accessed December 15, 2018.

21. In a paper proposing a new classification of the social and individual impact of various categories of drugs, psychedelics are at the very low end of the range and alcohol at the very high end (Nutt, King and Phillips 2010).

22. The Economist 2019.

23. Nour, Evans, and Carhart-Harris 2017.

24. The Inuit are the only indigenous people reported as not using psychedelics from plants in their social rituals, as their barren natural environment has none (Brody 2002).

25. Pollan, lecture at the John Adams Foundation in Amsterdam, December 11, 2018.

26. It is actually hardly a marginal phenomenon today. The U.S. National Institutes of Health estimate of U.S. psychedelic users is thirty million, or 20 percent of the population (Krebs and Johansen 2013).

27. Alexander 2017.

28. In his biographical notes Camus asks, "Peut-on, loin du sacré et de ses valeurs absolues, trouver la règle d'une conduite?" ("L'homme révolté," in *Cahiers 1945–1951*).

29. Kahneman 2013.

30. This treatment consists of a simplification of two different ideas in behavioral economics. The teapot example is the "endowment effect." That it gives more satisfaction not to lose a twenty-dollar bill than to find one is loss aversion. The endowment effect is seen as being based on loss aversion. However, this is not accepted by everyone. As a consequence, behavioral economists usually use them as distinct concepts.

31. Stoknes 2014.

Chapter 10. Beyond Incumbent Industry

Beinhocker summarizes the insights from Hannan and Freeman from a series of landmark studies on the "organizational ecology" of markets (Beinhocker 2006, p. 333). Extract from the book *The Origin of Wealth: Evolution, Complexity, and the Radical Remaking of Economics* by Eric D. Beinhocker. Published and used by permission of Harvard Business Publishing.

1. https://www.bp.com/en/global/corporate/sustainability/climate-change/carbon-pricing.html, accessed November 30, 2018.

2. The 2035 value of $55 is well below current global estimates of the social cost of carbon (SCC), but consistent with the current EPA value of $48. SCC values: median, US$417 per tonne of CO_2 (tCO_2); 66 percent confidence intervals, US$177–805 per tCO_2 (Ricke et al. 2018).

3. https://thehill.com/policy/energy-environment/415418-washington-state -voters-reject-carbon-tax, accessed November 30, 2018.

4. The total BP America contribution according to the Public Disclosure Commission was $13,125,362.77. Proponents of the measure raised less than half the sum. https://www.pdc.wa.gov/browse/campaign-explorer/committee?, accessed November 30, 2018.

5. Some the material in this chapter is adapted from an unpublished chapter originally intended for Colander and Kupers (2014).

6. Hsu 2019.

7. Barabási and Pósfai 2016.

8. The author worked at AT&T from 1988 to 1999.

9. This was not the cause of its ultimate demise and takeover by one of the former Baby Bells. The best explanation for this is found in the dependence on the high returns of the legacy long-distance business, which hampered expansion into new businesses. These high-growth businesses simply did not deliver the level of profits needed to sustain the corporate dividend policy. When this became increasingly clear to shareholders, the stock started to slide inexorably as a consequence. A similar dilemma awaits oil and gas companies as they attempt to diversify into lower-margin renewable and utility businesses—challenging the dividend policies that are the rock-hard foundation for their share price level.

10. Beinhocker 2006.

11. Brian Arthur, one of the very early complexity economists, describes how innovation is a recombination of diverse existing ideas. Companies are rewarded for specializing and thus naturally innovate in an evolutionary fashion by tweaking existing processes and products. More radical innovation is much more likely to occur outside existing firms (Arthur 2009).

12. See a treatment of resilience for the trade-off with efficiency, e.g., Walker and Salt (2006) or Kupers (2014).

13. https://www.sciencealert.com/these-100-companies-are-to-blame-for-71-of -the-world-s-greenhouse-gas-emissions, accessed November 15, 2018.

14. PricewaterhouseCooper Chief Purpose Officer Shannon Schuyler interview at https://www.foxbusiness.com/features/what-the-heck-is-a-chief-purpose -officer, accessed January 10, 2019.

15. Perez 2010.

16. De Geus 2002.

17. Krugman 1996.

18. "Airbus has announced that it's dropping plans to produce an E-Fan family of personal aircraft. It will move instead into developing a larger, more powerful aircraft, the E-Fan X, that could fly within three years." http://sustainableskies.org/airbus -e-fans-2-0-and-4-0-dropped-in-favor-of-e-fan-x/, accessed November 15, 2018.

19. "EasyJet is backing plans to develop commercial passenger aircraft powered by electric batteries instead of conventional aero engines. The prototype is going

to be developed by a new US firm called Wright Electric, which has already built a two-seat battery-powered plane." http://www.bbc.com/news/business -41404039, accessed November 15, 2018.
20. Mazzucato 2018.

Chapter 11. Who Pays for Learning?

1. While it appears that performance is uneven, there are actually real problems from lack of data to be sure (Kaplan and Lerner 2016).
2. Keynes 1923.
3. Homans 2012.
4. Wagner, Kåberger, et al. 2015.
5. Brady 2014.
6. Biello 2011.
7. Mazzucato 2014.
8. Jacobsen 2015.
9. Yergin 2011.
10. Van Benthem, Gillingham, and Sweeney 2008.
11. Newell, Jaffe, and Stavins 1998; Acemoglu, Aghion, et al. 2012.
12. Chancel and Piketty 2015.
13. Acemoglu, Aghion, et al. 2012; Newell, Jaffe, and Stavins 1998; Van Benthem, Gillingham, and Sweeney 2008.
14. Kojima and Koplow 2015; International Energy Agency 2014.
15. Iran is one of the few countries with massive fossil fuel subsidies to try to remove them. In 2010 it launched a reform program consisting of two components, a large price increase, ranging from fourfold for gasoline to ninefold for diesel, combined with generous cash transfers to compensate consumers. The program's difficulties illustrate how hard these reforms are, particularly in less than optimal governance contexts (Salehi-Isfahani 2017).
16. Wagner and Weitzman 2015; Kopp and Mignone 2012; Houser et al. 2015.
17. Wagner, Kåberger, et al. 2015.
18. Kavlak, McNerney, and Trancik 2018.
19. The EPA provides an overview of research into the ex ante and ex post cost of environmental regulation. The EU has looked into the same matter for its programs (U.S. EPA, OA, 2016; European Commission 2006).
20. Denny Ellerman 2003.
21. European Commission 2006.
22. Harrington, Morgenstern, and Nelson 2000; Hammitt 2000.
23. Arthur 2009.
24. Graziano and Gillingham 2014.

Chapter 12. Autonomous Cars as a Climate Policy

Thurner, Stefan, Rudolf Hanel, and Peter Klimek. 2018. *Introduction to the Theory of Complex Systems.* Oxford; New York: Oxford University Press p. 188. Reproduced with permission of Oxford Publishing Limited through PLSclear.

1. The percentage appears to be fairly consistent across cities. For an overview, see Barter 2013.
2. Assuming a standard four-floor building in its stead, the resulting eighty square meters would have a rental value of €24,000 per year, whereas the parking permits cost €2,000, or a charge of a mere 10 percent of the opportunity cost.
3. The European Commission lists 2016 road casualties as 25,600, excluding pollution impacts (European Commission 2018).
4. WHO 2015.
5. WHO 2018, vii.
6. Cowan and Hultén (1996) offer a first, comprehensive analysis of electric vehicle lock-in, in good part based on an earlier analysis of nuclear power plant design lock-in (Cowan 1990).
7. See Boeing (2017) for airplane fatality statistics. Human errors cause over 90 percent of car accidents. U.S. statistics for 2015 show a 93 percent rate (National Highway Traffic Safety Administration 2015).
8. A first fatal accident involving an autonomous vehicle occurred in Tempe, Arizona. This led to the suspension of the tests by Uber. The cause is under investigation, but is likely to be difficult to pinpoint with precision, given the self-learning nature of the algorithms (NTSB 2018). Like every accident, this is a tragedy, but overall indications are that the dismal safety record of cars with drivers can be vastly improved through autonomous technology.
9. Field and Clark 1997.
10. There is considerable speculation and study of the impact of autonomous vehicles on trucking employment. It ranges from a small number who project growth in employment (UberATG 2018) to those who project mass loss of jobs, such as Yang 2018.
11. This general phenomenon often comes under the heading of the "rebound effect" or the "Jevons Paradox"—more accurately "Jevons's Paradox." See York and McGee 2016 for an overview.
12. For a good survey of this decades-old debate, see *The Wizard and the Prophet* (C. Mann 2018).
13. Bertoncello and Wee 2015.
14. B. Smith 2016.
15. "Such vehicle systems—even though they may possibly be not overrideable at any time or even though they may not be switched off completely—may help the driver to maintain his vehicle under control in dangerous driving situations" (ECE 2014).
16. Germany, with parts of its highways without speed limits, is a significant exception, given its large global car industry.
17. http://www.aprilli.com/autonomous-travel-suite/, accessed November 5, 2018.
18. The true safety of robotic cars will not become fact until sufficient statistical evidence has been assembled. It is, however, possible to infer radically improved safety from first principles. For example, an essential difference with robotic cars is the ability to learn from all mistakes, not only from one's individual ones, as with a human driver. National Highway Traffic Safety Administration chief Mark Rosekind stated it as follows: "If we wait for perfect, we'll be waiting for a very, very long time. How many lives might we be losing while we wait?"

Rosekind said that self-driving cars will be able to improve even as they make mistakes. Data from an accident involving a self-driving car "can be taken, analyzed, and then the lessons can be shared" with manufacturers of all automated vehicles. Human drivers "must learn on the road and make the same mistakes as thousands before them." Quoted in Carroll and Shepardson (2016).

19. Albert and McCaig 2015.
20. Moor 2016.
21. It is becoming increasingly accepted that cars contribute to a lethal degradation of air quality in cities. The fine particles coming out of the tailpipes pretty much double the fatalities from road accidents, not even counting diesel fraud. "The largest contributors for both pollutant-related mortalities are road transportation, causing ~53,000 (90% CI: 24,000–95,000) $PM_{2.5}$-related deaths and ~5000 (90% CI: –900 to 11,000) ozone-related early deaths per year, and power generation, causing ~52,000 (90% CI: 23,000–94,000) $PM_{2.5}$-related and ~2000 (90% CI: –300 to 4000) ozone-related premature mortalities per year. Industrial emissions contribute to ~41,000 (90% CI: 18,000–74,000) early deaths from $PM_{2.5}$ and ~2000 (90% CI: 0–4000) early deaths from ozone" (Caiazzo et al. 2013).
22. "From World War II until just a few years ago, the number of miles driven annually on America's roads steadily increased. Then, at the turn of the century, something changed: Americans began driving less. By 2011, the average American was driving 6 percent fewer miles per year than in 2004." See B. Davis, Dutzik, and Baxandall (2012).
23. This sounds hard, but in fact from 1996 to 2006, every state enacted Graduated Drivers' Licensing (GDL) laws. GDL laws, which are designed to keep young people safe, also make obtaining a driver's license more challenging (ibid.).
24. In December 2011 the National Transportation Safety Board recommended banning cell phone use while driving entirely (ibid.).
25. E.g., Royal Dutch Shell, where the author was employed from 1998 to2010.
26. From a local blog found at http://www.boomchicago.nl/amsterdamblog/using-uber-pop/, accessed September 20, 2018.

Conclusion: Beyond a Blueprint for the Climate Revolution

1. Blom (2019) provides a sweeping account of the consequences of global cooling, with the danger of attributing to it all the historical changes of the seventeenth century. This illustrates the challenge of causality in complex systems. A twenty-first century example concerns the Syrian refugee stream into Europe in 2015 and 2016. This is widely understood to be in good part triggered by climate-change droughts and has had widespread consequences. Attributing causality for the rise of the anti-immigrant vote in Europe to climate change is wrong in the sense of direct causality, but clearly it has been a major contributing factor, among others. The same applies to seventeenth-century events.
2. We highlight the influence of economics on policy as determinant in practice. The reflection on what influences the dynamics of the socioeconomic system is

much broader in other social disciplines. Merton (1936) is an example of an early attempt at characterizing the dynamics of policy in terms of complexity dynamics.

3. Lindblom 1959.

4. KPMG 2018.

5. An additional policy argument in Singapore is that because of the lack of renewable resources in the country, with an exclusively fossil fuel–based power system, a shift to electric cars would only displace emissions and not reduce them.

6. In Donald Rumsfeld's famous quote, he actually excluded this option, listing only the three other permutations of known and unknown. He could have avoided overlooking this relevant option for the Iraq War justification if he had gone back to the source, the fourteenth-century Sufi Persian poet Ibn Yamin:

> One who knows and knows that he knows . . .
> His horse of wisdom will reach the skies.
> One who knows, but doesn't know that he knows . . .
> He is fast asleep, so you should wake him up!
> One who doesn't know, but knows that he doesn't know . . .
> His limping mule will eventually get him home.
> One who doesn't know and doesn't know that he doesn't know . . .
> He will be eternally lost in his hopeless oblivion!

7. Wired 2015.

8. The Louvre dedicated an exhibition to this topic, illustrating the phenomenon with examples from ancient Egypt to modern France (Théâtre du Pouvoir—Petite Galerie du Louvre, September 27, 2017–July 2, 2018).

9. Davies (2017) relates the story of the DARPA challenge. This race was preceded by another one more than a century earlier, when *Le Petit Journal* organized a Paris-Rouen race for horseless carriages in December 1893. The 102 entrants had a mix of engines, powered by electricity, compressed air, and petrol. The distance between Paris and Rouen doomed the electric engines, which needed a battery swap every thirty kilometers, which was not allowed under the rules of the race. The clear winner was the internal combustion engine (The Economist 2017b).

10. More precisely, the goal provides a common parameter that influences the individual agent behavior and the micro level. As such it provides a coordination mechanism that influences the emergent behavior of the system.

11. The RSPO has developed a set of environmental and social criteria that companies must comply with in order to produce certified sustainable palm oil (CSPO). See http://www.rspo.org.

12. When challenged, companies recognized that RSPO targets fall short. One solution is the RSPO NEXT initiative. The fact that it brings in basics such as not burning forests, managing greenhouse gas emissions, etc., is a sign of how deficient the basic structure is: "The components of RSPO NEXT fall into the following categories: No Deforestation, No Fire, No Planting on Peat, Reduction of GHGs, Respect for Human Rights and Transparency and are applicable

at an organization wide level, including investments, joint ventures and in the organization's wider supply base" (www.rspo.org/certification/rspo-next).

13. Jaeger and Jaeger 2011.

14. This is consistent with Elinor Ostrom's recommendation for a more polycentric approach. In Ostrom's perspective, shared information and joint goals function as an essential ingredient for collective action. A ratcheting mechanism itself builds a process for ownership of the information. See Ostrom (2009).

15. https://www.ey.com/en_gl/power-utilities/is-any-end-in-sight-for-power-and-utilities-asset-impairments-in-europe, accessed January 14, 2019.

16. http://www.chinadaily.com.cn/china/19thcpcnationalcongress/2017-11/04/content_34115212.htm, accessed January 14, 2019.

17. https://ec.europa.eu/clima/policies/strategies/2050_en, accessed January 14, 2019.

18. Beede, Powers, and Ingram 2017.

19. Berger, Chen, and Frey 2018.

20. The Kinder Institute at Rice University does an annual survey of Houston, recently complemented by a similar study of Copenhagen. The comparison highlights some of the underlying drivers for its urban characteristics, notably the role of trust. For comparison of the reports, see Holeywell (2015).

21. That trust is an essential component of a well-functioning society is not news for social scientists. Complexity scientists have tried to more precisely model its role through simulations. For our purposes, we highlight trust as a necessary ingredient to allow the kinds of network dynamics to develop that are essential precursors for phase transitions. See, for example, Chen, Chie, and Zhang (2015).

Epilogue

Fifteen-year-old Greta Thurnberg of Sweden gave a keynote speech at COP24 in Katowice, Poland, flanked by the UN secretary general.

1. Rigg and Mason (2018) point to the so-called shallow interdisciplinarity among the natural sciences that dominates climate science, where data that can be collected through quantitative scientific methods is prioritized, leading to standardized and linear policy prescriptions. It is contrasted with "deep interdisciplinarity" (working across the natural and social sciences and humanities), which is far less common and far more challenging.

2. It's actually probably not strictly Goethe: "Although it was most probably composed by Georg Tobler between his visits to Weimar in 1781 and 1782, this prose rhapsody nevertheless reflects Goethe's conception of nature. It was included in Goethe's works, and until the end of the 19th century, Goethe was assumed to be its author" (Fullenwider 1986).

3. For example, Sterner et al. (2019) emphasize the interconnectedness between the planetary boundaries.

4. Quoted in an *Economist* article: "The moral assumptions embedded in economic models of climate change" (The Economist 2018b).

5. See, for example, Nordhaus's review of the Stern report (Nordhaus 2007).

6. Epstein 2008.

REFERENCES

Acemoglu, Daron, Philippe Aghion, Leonardo Bursztyn, and David Hemous.
    2012. "The Environment and Directed Technical Change." *American
    Economic Review* 102 (1): 131–66. https://doi.org/10.1257/aer.102.1.131.
Acemoglu, Daron, Ufuk Akcigit, Douglas Hanley, and William Kerr. 2016.
    "Transition to Clean Technology." *Journal of Political Economy* 124 (1).
    https://ideas.repec.org/a/ucp/jpolec/doi10.1086-684511.html.
Agentur für Erneuerbare Energien. 2017. "Grafik-Dossier Akzeptanzumfrage
    2017." 2017. https://www.unendlich-viel-energie.de/mediathek/grafiken
    /grafik-dossier-akzeptanzumfrage-2017.
Aghion, Philippe, Cameron Hepburn, Alexander Teytelboym, and Dimitri
    Zenghelis. 2014. "Path Dependence, Innovation and the Economics of
    Climate Change." A contributing paper from Centre for Climate Change
    Economics and Policy and Grantham Research Institute on Climate Change
    and the Environment to The New Climate Economy. The Global Commis-
    sion on the Economy and Climate. https://newclimateeconomy.report
    /workingpapers/wp-content/uploads/sites/5/2016/04/Path-dependence-and
    -econ-of-change.pdf
Akerlof, George A. 2007. "The Missing Motivation in Macroeconomics."
    American Economic Review 97 (1): 32.
Alberici, Sacha, Sil Boeve, Pieter van Breevoort, Yvonne Deng, Sonja Förster,
    Valentijn van Gastel, Katharina Grave, et al. 2014. "Subsidies and Costs of
    EU Energy." European Commission. https://ec.europa.eu/energy/sites/ener
    /files/documents/ECOFYS%202014%20Subsidies%20and%20costs%20
    of%20EU%20energy_11_Nov.pdf.
Albert, Michael, and Lind F. McCaig. 2015. "Emergency Department Visits for
    Motor Vehicle Traffic Injuries: United States, 2010–2011." NCHS Data Brief
    185. Centers for Disease Control.
Alexander, Frederick H. 2017. "Next Phase for Benefit Corporation Governance
    Begins." *Westlaw Journal Corporate Officers and Directors Liability*,

December. http://www.mnat.com/files/BylinedArticles/2018_WLJ_COD3312 _RickAlexander.pdf.

Ansar, Antif, Ben Caldecott, and James Tilbury. 2013. "Stranded Assets and the Fossil Fuel Divestment Campaign: What Does Divestment Mean for the Valuation of Fossil Fuel Assets?" Stranded Assets Programme. Oxford, UK: Smith School of Enterprise and the Environment, University of Oxford. http://www.smithschool.ox.ac.uk/publications/reports/SAP-divestment-report -final.pdf.

Arthur, W. Brian. 1989. "Competing Technologies, Increasing Returns, and Lock-In by Historical Events." *Economic Journal* 99 (394): 116–31. https://doi.org/10.2307/2234208.

———. 1994. *Increasing Returns and Path Dependence in the Economy.* Ann Arbor: University of Michigan Press.

———. 2009. *The Nature of Technology: What It Is and How It Evolves.* New York: Simon and Schuster.

———. 2014. *Complexity and the Economy.* Oxford: Oxford University Press.

———. 2019 *Kenneth Arrow and nonequilibrium economics*, Quantitative Finance, 19:1, 29–31, DOI: 10.1080/14697688.2018.1533738.

Axelrod, Robert, and William D. Hamilton. 1981. "The Evolution of Cooperation." *Science* 211 (4489): 1390–96.

Baake, Rainer. 2013. "12 Insights on Germany's Energiewende." Agora Energiewende. https://www.agora-energiewende.de/fileadmin/Projekte/2012/12 -Thesen/Agora_12_Insights_on_Germanys_Energiewende_web.pdf.

Balint, T., F. Lamperti, A. Mandel, M. Napoletano, A. Roventini, and A. Sapio. 2017. "Complexity and the Economics of Climate Change: A Survey and a Look Forward." *Ecological Economics* 138 (August): 252–65. https://doi.org /10.1016/j.ecolecon.2017.03.032.

Ball, Jeffrey. 2018. "Why Carbon Pricing Isn't Working—Good Idea in Theory, Failing in Practice." *Foreign Affairs*, August. https://www.foreignaffairs.com /articles/world/2018-06-14/why-carbon-pricing-isnt-working.

Barabási, Albert-László. 2016. *Network Science.* Cambridge: Cambridge University Press. http://networksciencebook.com

Barter, Paul. 2013. "'Cars Are Parked 95% of the Time.' Let's Check!" *Reinventing Parking* (blog), February 22. https://www.reinventingparking.org /2013/02/cars-are-parked-95-of-time-lets-check.html.

BBC. 2014. *How Reintroducing Wolves Helped Save a Famous Park.* BBC. http://www.bbc.com/future/story/20140128-how-wolves-saved-a-famous-park.

Beede, David, Regina Powers, and Cassandra Ingram. 2017. "The Employment Impact of Autonomous Vehicles." ESA Issue Brief #05-17. U.S. Department of Commerce. http://www.esa.doc.gov/sites/default/files/Employment%20 Impact%20Autonomous%20Vehicles_0.pdf.

Beinhocker, Eric D. 2006. *The Origin of Wealth: Evolution, Complexity, and the Radical Remaking of Economics.* Boston: Harvard Business Press.

Berger, Thor, Chinchih Chen, and Carl Benedikt Frey. 2018. "Drivers of Disruption? Estimating the Uber Effect." *European Economic Review* 110 (November): 197–210. https://doi.org/10.1016/j.euroecorev.2018.05.006.

Bertoncello, Michele, and Dominik Wee. 2015. "Ten Ways Autonomous Driving Could Redefine the Automotive World." McKinsey. June. https://www .mckinsey.com/industries/automotive-and-assembly/our-insights/ten-ways -autonomous-driving-could-redefine-the-automotive-world.

Beveridge, Ross, and Kristine Kern. 2013. "The Energiewende in Germany: Background, Developments and Future Challenges." *Renewable Energy Law and Policy Review (RELP)* 4: 3.

Biello, David. 2011. "How Solyndra's Failure Promises a Brighter Future for Solar Power." *Scientific American*, October 12. https://www.scientificamerican.com /article/how-solyndras-failure-helps-future-of-solar-power/.

Blanchard, Olivier. 2000. "What Do We Know about Macroeconomics That Fisher and Wicksell Did Not?" *De Economist* 148 (5): 571–601. https://doi .org/10.1023/A:1004131616369.

Blom, Philipp. 2019. *Nature's Mutiny: How the Little Ice Age of the Long Seventeenth Century Transformed the West and Shaped the Present.* New York: Liveright.

Boeing. 2017. "Effect of Reducing Maintenance Errors." *AERO*. http://www .boeing.com/commercial/aeromagazine/articles/qtr_2_07/AERO_Q207.pdf.

Boyce, Mark S. 2018. "Wolves for Yellowstone: Dynamics in Time and Space." *Journal of Mammalogy* 99 (5): 1021–31. https://doi.org/10.1093/jmammal /gyy115.

Brady, Jeff. 2014. "After Solyndra Loss, U.S. Energy Loan Program Turning a Profit." NPR.Org, November 13. http://www.npr.org/2014/11/13 /363572151/after-solyndra-loss-u-s-energy-loan-program-turning-a-profit.

Brody, Hugh. 2002. *The Other Side of Eden: Hunters, Farmers, and the Shaping of the World.* New York: North Point Press.

Broido, Anna D., and Aaron Clauset. 2018. "Scale-Free Networks Are Rare." *ArXiv Preprint ArXiv:1801.03400.*

Buckley, Walter. 1968. "Society as a Complex Adaptive System." In *Modern Systems Research for the Behavioral Scientist*, edited by Walter Buckley. Chicago: Aldine.

Bundesministerium für Wirtschaft und Energie (BMWi). 2019. "Kommission 'Wachstum, Strukturwandel und Beschäftigung' Abschuussbericht." Berlin. https://www.bmwi.de/Redaktion/DE/Downloads/A/abschlussbericht -kommission-wachstum-strukturwandel-und-beschaeftigung.pdf?__blob =publicationFile.

Butt, Ronald. 1981. "Mrs. Thatcher: The First Two Years," *Sunday Times*, March 5. https://www.margaretthatcher.org/document/104475.

Caiazzo, Fabio, Akshay Ashok, Ian A. Waitz, Steve H. L. Yim, and Steven R. H. Barrett. 2013. "Air Pollution and Early Deaths in the United States. Part I: Quantifying the Impact of Major Sectors in 2005." *Atmospheric Environment* 79 (November): 198–208. https://doi.org/10.1016/j.atmosenv.2013.05.081.

Cairney, Paul, and Robert Geyer. 2017. "A Critical Discussion of Complexity Theory: How Does 'Complexity Thinking' Improve Our Understanding of Politics and Policymaking?" *Complexity, Governance and Networks* 3 (2): 1–11. https://doi.org/10.20377/cgn-56.

Carroll, Rory, and David Shepardson. 2016. "Top U.S. Vehicle Safety Regulator Stands by Self-Driving Cars." Reuters, July 20. https://www.reuters.com /article/us-autos-selfdriving/top-u-s-vehicle-safety-regulator-stands-by-self -driving-cars-idUSKCN1002U7.

Chancel, Lucas, and Thomas Piketty. 2015. "Carbon and Inequality: From Kyoto to Paris—Trends in the Global Inequality of Carbon Emissions (1998–2013) and Prospects for an Equitable Adaptation Fund." Paris School of Economics. http://piketty.pse.ens.fr/files/ChancelPiketty2015.pdf.

Chen, Shu-Heng, Bin-Tzong Chie, and Tong Zhang. 2015. "Network-Based Trust Games: An Agent-Based Model." *Journal of Artificial Societies and Social Simulation* 18 (3): 5.

Christakis, Nicholas A. 2007. "The Spread of Obesity in a Large Social Network over 32 Years." *New England Journal of Medicine* 357: 370–379.

Coates, Ben. 2015. "How Green Is Holland? From Carbon Emissions to Climate Change." *The Independent*, September 21. https://www.independent.co.uk /environment/green-living/how-green-is-holland-from-carbon-emissions-to -climate-change-10511649.html.

Cohn, Alain, Michel André Maréchal, David Tannenbaum, and Christian Lukas Zünd. 2019. "Civic Honesty around the Globe." *Science* Vol. 365, Issue 6448, pp. 70–73. https://doi.org/10.1126/science.aau8712.

Colander, David, Richard P. F. Holt, and J. Barkley Rosser. 2004. *The Changing Face of Economics: Conversations with Cutting Edge Economists*. Ann Arbor: University of Michigan Press.

Colander, David, and Roland Kupers. 2014. "Complexity and the Art of Public Policy." Princeton University Press. https://press.princeton.edu/titles/10207.html.

Colander, David, Roland Kupers, Thomas Lux, and Casey Rothschild. 2010. "Reintegrating the Social Sciences." In Middlebury College Economics Discussion Paper No. 10–33:10. http://sandcat.middlebury.edu/econ/repec /mdl/ancoec/1033.pdf.

Council of Economic Advisers. 2017. "Economic Report of the President." Washington, DC. https://www.govinfo.gov/features/ERP-2017.

Cowan, Robin. 1990. "Nuclear Power Reactors: A Study in Technological Lock-In." *Journal of Economic History* 50 (3): 541–67. https://doi.org/10 .1017/S0022050700037153.

Cowan, Robin, and Staffan Hultén. 1996. "Escaping Lock-in: The Case of the Electric Vehicle." In "Technology and the Environment," special issue, *Technological Forecasting and Social Change* 53 (1): 61–79. https://doi.org /10.1016/0040-1625(96)00059-5.

Creutzig, Felix, Joyashree Roy, William F. Lamb, Inês M. L. Azevedo, Wändi Bruine de Bruin, Holger Dalkmann, Oreane Y. Edelenbosch, et al. 2018. "Towards Demand-Side Solutions for Mitigating Climate Change." *Nature Climate Change* 8 (4): 260–63. https://doi.org/10.1038/s41558-018-0121-1.

David, Paul A. 1985. "Clio and the Economics of QWERTY." *American Economic Review* 75 (2): 332–37.

Davies, Alan. 2017. "An Oral History of the 2004 Darpa Grand Challenge, the Grueling Robot Race That Launched the Self-Driving Car." *Wired*,

August 3. https://www.wired.com/story/darpa-grand-challenge-2004-oral -history/.

Davis, Benjamin, Tony Dutzik, and Phineas Baxandall. 2012. "Transportation and the New Generation—Why Young People Are Driving Less and What It Means for Transportation Policy." Frontier Group/US PIRG Education Group. https://frontiergroup.org/reports/fg/transportation-and-new-generation.

Davis, Mark. 1996. *One River: Explorations and Discoveries in the Amazon Rain Forest*. New York: Simon and Schuster.

De Geus, A. 2002. *The Living Company: Growth, Learning and Longevity in Business*. Brighton: Harvard Business Review Press.

Denny Ellerman, A. 2003. "Ex Post Evaluation of Tradable Permits: The U.S. SO$_2$ Cap-and-Trade Program." 03-003 WP. MIT Center for Energy and Environmental Policy Research. https://pdfs.semanticscholar.org/dc0d/8b481 397d87b8665e8ea89b88bcec9b0cf97.pdf.

Der Spiegel. 2011. "Out of Control: Merkel Gambles Credibility with Nuclear U-Turn." Spiegel Online, March 21. http://www.spiegel.de/international/germany /out-of-control-merkel-gambles-credibility-with-nuclear-u-turn-a-752163.html.

Deutsche Welle. 2019. "Merkel Will CO2-Neutralität Bis 2050." Deutsche Welle. https://www.dw.com/de/merkel-will-co2-neutralit%C3%A4t-bis-2050/a -48730300.

Dunne, Jennifer A., Richard J. Williams, and Neo D. Martinez. 2002. "Food-Web Structure and Network Theory: The Role of Connectance and Size." *Proceedings of the National Academy of Sciences* 99 (20): 12917–22. https://doi.org/10.1073/pnas.192407699.

Dusza, Karl. 1989. "Max Weber's Conception of the State." *International Journal of Politics, Culture, and Society* 3 (1): 71–105.

ECE. 2014. *Report of the Sixty-Eighth Session of the Working Party on Road Traffic Safety*. ECE/TRANS/WP.1/145. UN Economic and Social Council— Economic Commission for Europe Inland Transport Committee Working Party on Road Traffic Safety. http://www.unece.org/fileadmin/DAM/trans/doc /2014/wp1/ECE-TRANS-WP1-145e.pdf.

The Economist. 1913. "Neighbours and Friends." June 28.

———. 2007. "What Is This That Roareth Thus?." September 6. http://www .economist.com/node/9719105.

———. 2013. "How to Lose Half a Trillion Euros." October 15. https://www .economist.com/briefing/2013/10/15/how-to-lose-half-a-trillion-euros.

———. 2017a. "Crime and Despair in Baltimore." June 29. https://www .economist.com/united-states/2017/06/29/crime-and-despair-in-baltimore.

———. 2017b. "The Death of the Internal Combustion Engine." December 8. https://www.economist.com/leaders/2017/08/12/the-death-of-the-internal -combustion-engine.

———. 2018a. "Local Experiments with Reform Are Becoming Rarer under Xi Jinping." August 18. https://www.economist.com/china/2018/08/18/local -experiments-with-reform-are-becoming-rarer-under-xi-jinping.

———. 2018b. "The Moral Assumptions Embedded in Economic Models of Climate Change." December 6. https://www.economist.com/finance-and

-economics/2018/12/08/the-moral-assumptions-embedded-in-economic
-models-of-climate-change.

———. 2018c. "Why People in Rich Countries Are Eating More Vegan Food."
The Economist, October 13. https://www.economist.com/briefing/2018/10
/13/why-people-in-rich-countries-are-eating-more-vegan-food.

———. 2019. "Daily chart: What is the most dangerous drug?" *The Economist*,
June 25. https://www.economist.com/graphic-detail/2019/06/25/what-is-the
-most-dangerous-drug.

Edlin, Aaron S., and Pinar Karaca-Mandic. 2006. "The Accident Externality
from Driving." *Journal of Political Economy* 114 (5): 931–55. https://doi.org
/10.1086/508030.

Energy Transitions Commission. 2018. "Mission Possible—Reaching Net-Zero
Carbon Emissions from Harder-to-Abate Sectors by Mid-Century." Energy
Transitions Commission. http://www.energy-transitions.org/sites/default/files
/ETC_MissionPossible_FullReport.pdf.

Epstein, Joshua M. 2008. "Why Model?" *Journal of Artificial Societies and
Social Simulation* 11, (4): 12.

European Commission. 2006. "Ex-Post Costs to Business of EU Environmental
Legislation." Commissioned by European Commission. Institute for
Environmental Studies, Vrije Universiteit. http://ec.europa.eu/environment
/enveco/ex_post/pdf/costs.pdf.

———. 2018. "Road Fatalities in the EU since 2001." *Mobility and Transport*.
May 24. /transport/road_safety/specialist/statistics_en.

Evans, Simon. 2015. "Climate Showdown: Has the US, UK or Germany Done
More to Cut Emissions?" *Carbon Brief*, April 10. https://www.carbonbrief.org
/climate-showdown-has-the-us-uk-or-germany-done-more-to-cut-emissions.

Farmer, J. D., C. Hepburn, M. C. Ives, T. Hale, T. Wetzer, P. Mealy, R. Rafaty, S.
Srivastav, and R. Way. 2019. "Sensitive Intervention Points in the Post-
Carbon Transition." *Science* 364 (6436): 132–34. https://doi.org/10.1126
/science.aaw7287.

Feng, Kuishuang, Steven J. Davis, Laixiang Sun, and Klaus Hubacek. 2015.
"Drivers of the US CO_2 Emissions 1997–2013." *Nature Communications* 6
(July): 7714. https://doi.org/10.1038/ncomms8714.

———. 2016. "Correspondence: Reply to 'Reassessing the Contribution of
Natural Gas to US CO_2 Emission Reductions since 2007.'" *Nature Communications* 7 (March): 10693. https://doi.org/10.1038/ncomms10693.

Field, Frank R., III, and Joel P. Clark. 1997. "A Practical Road to Lightweight
Cars—MIT Technology Review." *MIT Technology Review*, January 1,
1997. https://www.technologyreview.com/s/400002/a-practical-road-to
-lightweight-cars/.

Florini, Ann. 1996. "Evolution of International Norms." *International Studies
Quarterly* 40 (3): 363–89. https://doi.org/10.2307/2600716.

Fouquet, Roger. 2016. "Path Dependence in Energy Systems and Economic
Development." *Nature Energy* 1 (8): 16098.

Foust, Christina R., and William O'Shannon Murphy. 2009. "Revealing
and Reframing Apocalyptic Tragedy in Global Warming Discourse."

Environmental Communication 3 (2): 151–67. https://doi.org/10.1080 /17524030902916624.

Fowler, James H., Timothy R. Johnson, James F. Spriggs, Sangick Jeon, and Paul J. Wahlbeck. 2007. "Network Analysis and the Law: Measuring the Legal Importance of Precedents at the U.S. Supreme Court." *Political Analysis* 15 (3): 324–46. https://doi.org/10.1093/pan/mpm011.

Fullenwider, Henry. 1986. "The Goethean Fragment 'Die Natur' in English Translation." *Comparative Literature Studies* 23 (2): 170–77.

Gao, Jian, Yi-Cheng Zhang, and Tao Zhou. 2019. "Computational Socio-economics." *ArXiv:1905.06166 [Physics, q-Fin]*, May. http://arxiv.org/abs /1905.06166.

Gardner, Martin. 2005. *The New Ambidextrous Universe: Symmetry and Asymmetry from Mirror Reflections to Superstrings*. Third rev. ed. Mineola, NY: Dover Publications.

Garfield, Michael. 2018. "Thinking Interplanetary: A Conversation with David Krakauer, President of the Santa Fe Institute." *Medium* (blog), July 17. https://medium.com/@michaelgarfield/thinking-interplanetary-a-conversation -with-david-krakauer-president-of-the-santa-fe-institute-371cc46af542.

Gell-Mann, Murray. 1995. *The Quark and the Jaguar: Adventures in the Simple and the Complex*. New York: Macmillan.

Gladwell, Malcolm. 2000. *The Tipping Point: How Little Things Can Make a Big Difference*. London: Abacus.

Goodell, Jeff. 2010. *How to Cool the Planet: Geoengineering and the Audacious Quest to Fix Earth's Climate*. Boston: Houghton Mifflin Harcourt.

———. 2017. *The Water Will Come: Rising Seas, Sinking Cities, and the Remaking of the Civilized World*. Boston: Little, Brown and Company.

Graichen, Patrick, and Thorsten Lenck. 2017. "Energiepreise und Ener-giewende," https://www.agora-energiewende.de/fileadmin2/Projekte/2017 /Abgaben_Umlagen/Foliensatz_Abgaben-Umlagen_Grundlagen_2017-04-10 .pdf.

Grave, Katharina, Felix von Blücher, Barbara Breitschopf, and Martin Pudlik. 2015. "Strommärkte im internationalen Vergleich." Fraunhofer ISI—Ecofys. https://www.isi.fraunhofer.de/content/dam/isi/dokumente/ccx/2015 /Industriestrompreise_Strommaerkte.pdf.

Graziano, Marcello, and Kenneth Gillingham. 2014. "Spatial Patterns of Solar Photovoltaic System Adoption: The Influence of Neighbors and the Built Environment." *Journal of Economic Geography* 15 (4): 815–39. https://doi .org/10.1093/jeg/lbu036.

Groot, A. de; Delft, Y.C. van. 2018. "A first order roadmap for Electrification of the Dutch Industry." ECN Biomass & Energy Efficiency ECN-O—18-002 https://publicaties.ecn.nl/PdfFetch.aspx?nr=ECN-O--18-002, accessed on January 21, 2019.

Gross, Robert, Richard Hanna, Ajay Gambhir, Philip Heptonstall, and Jamie Speirs. 2018. "How Long Does Innovation and Commercialisation in the Energy Sectors Take? Historical Case Studies of the Timescale from Inven-tion to Widespread Commercialisation in Energy Supply and End Use

Technology." *Energy Policy* 123 (December): 682–99. https://doi.org/10.1016/j.enpol.2018.08.061.

Hamblin, James. 2017. "Substituting Beans for Beef Would Help the U.S. Meet Climate Goals." *The Atlantic*, February 8.

Hamer, Mick. 2017. *A Most Deliberate Swindle: How Edwardian Fraudsters Pulled the Plug on the Electric Bus and Left Our Cities Gasping for Breath.* Brentford: RedDoor.

Hammitt, James K. 2000. "Are the Costs of Proposed Environmental Regulations Overestimated? Evidence from the CFC Phaseout." *Environmental and Resource Economics* 16 (3): 281–302. https://doi.org/10.1023/A:1008352022368.

Hardin, Garrett. 1968. "The Tragedy of the Commons." *Science* 162 (3859): 1243–48. https://doi.org/10.1126/science.162.3859.1243.

Harrington, Winston, Richard D. Morgenstern, and Peter Nelson. 2000. "On the Accuracy of Regulatory Cost Estimates." *Journal of Policy Analysis and Management* 19 (2): 297–322. https://doi.org/10.1002/(SICI)1520-6688(200021)19:2<297::AID-PAM7>3.0.CO;2-X.

Hathaway, Oona A., and Scott J. Shapiro. 2017. *The Internationalists: How a Radical Plan to Outlaw War Remade the World.* New York: Simon and Schuster.

Heal, Geoffrey. 2017. "The Economics of the Climate." *Journal of Economic Literature* 55 (3): 1046–63.

Holeywell, Ryan. 2015. "What Copenhagen and Houston Tell Us about How 'Trust' Affects Cities." Kinder Institute for Urban Research, June 10. https://kinder.rice.edu/2015/10/06/copenhagen-and-houston.

Homans, Charles. 2012. "The Experiment." *New Republic*, January 25. https://newrepublic.com/article/100037/steven-chu-energy-obama-solyndra.

Houser, Trevor, Solomon Hsiang, Robert Kopp, and Kate Larsen. 2015. *Economic Risks of Climate Change: An American Prospectus.* New York: Columbia University Press.

Hsu, Tiffany. 2019. "The World's Last Blockbuster Has No Plans to Close." *New York Times*, March 6. https://www.nytimes.com/2019/03/06/business/last-blockbuster-store.html.

International Energy Agency. 2014. "World Energy Outlook." https://www.iea.org/publications/freepublications/publication/WEO2014.pdf.

IPCC. 1992. *First Assessment Report.*

———. 1995. *Second Assessment Report.*

———. 2001. *Third Assessment Report.*

———. 2007. *Fourth Assessment Report.*

———. 2013. "*Fifth Assessment Report: Climate Change.*" http://www.ipcc.ch/report/ar5/wg1/.

———. 2018. *Special Report—Global Warming of 1.5°C.* Summary for Policy-makers IPCC SR1.5. http://report.ipcc.ch/sr15/pdf/sr15_spm_final.pdf.

Jackson, Tim. 2016. *Prosperity without Growth: Foundations for the Economy of Tomorrow.* 2nd ed. Abingdon UK: Routledge.

Jacobs, Jane. 2016. *The Death and Life of Great American Cities.* New York: Knopf Doubleday Publishing Group.

Jacobsen, Annie. 2015. *The Pentagon's Brain: An Uncensored History of DARPA, America's Top-Secret Military Research Agency*. Boston: Little, Brown and Company. Jaeger, Carlo, Leonidas Paroussos, Diana Mangalagiu, Roland Kupers, Antoine Mandel, and Joan David Tabara. 2011. *A New Growth Path for Europe: Generating Prosperity and Jobs—Synthesis Report*. Study commissioned by the German Federal Ministry for the Environment. https://www.researchgate.net/publication/254761519_A_new _growth_path_for_Europe_Generating_prosperity_and_jobs.

Jaeger, Carlo C., and Julia Jaeger. 2011. "Three Views of Two Degrees." *Regional Environmental Change* 11 (1): 15–26. https://doi.org/10.1007/s10113-010 -0190-9.

Jun, Ma, and Simon Zadek. 2019. "Decarbonizing the Belt and Road: A Green Finance Roadmap." A collaboration between the Tsinghua University Center for Finance and Development, Vivid Economics and the Climateworks Foundation. https://www.vivideconomics.com/casestudy/decarbonizing-the -belt-and-road-initiative-a-green-finance-roadmap/.

Kahneman, Daniel. 2013. *Thinking, Fast and Slow*. New York: Farrar, Straus and Giroux.

Kaplan, Steven, and Josh Lerner. 2016. "Venture Capital Data: Opportunities and Challenges." Working Paper 22500 Cambridge, MA: National Bureau of Economic Research. https://doi.org/10.3386/w22500.

Kavlak, Goksin, James McNerney, and Jessika E. Trancik. 2018. "Evaluating the Causes of Cost Reduction in Photovoltaic Modules." *Energy Policy* 123 (December): 700–710. https://doi.org/10.1016/j.enpol.2018.08.015.

Keith, David. 2013. *A Case for Climate Engineering*. Cambridge, MA: MIT Press.

Keith, David W. 2000. "Geoengineering the Climate: History and Prospect." *Annual Review of Energy and the Environment* 25 (1): 245–84. https://doi .org/10.1146/annurev.energy.25.1.245.

Keynes, John M. 1923. *A Tract on Monetary Reform*. New York: Macmillan.

Keynes, John M. 2010. *Economic Possibilities for Our Grandchildren*. In: Essays in Persuasion. London: Palgrave Macmillan.

Kirman, Alan P. 1992. "Whom or What Does the Representative Individual Represent?" *Journal of Economic Perspectives* 6 (2): 117–36. https://doi.org /10.1257/jep.6.2.117.

Klarreich, Erica. 2018. "Scant Evidence of Power Laws Found in Real-World Networks." *Quanta Magazine*, February 15. https://www.quantamagazine .org/scant-evidence-of-power-laws-found-in-real-world-networks-20180215/.

Klein, Naomi. 2011. "Capitalism vs. the Climate." *The Nation*, November 9. https://www.thenation.com/article/capitalism-vs-climate/.

Klimaatakkoord. 2018. "Klimaatberaad—Over het Klimaatakkoord— Klimaatakkoord." Webpagina, April 18. https://www.klimaatakkoord.nl /klimaatakkoord/klimaatberaad.

Kojima, Masami, and Doug Koplow. 2015. "Fossil Fuel Subsidies: Approaches and Valuation." SSRN Scholarly Paper ID 2584245. Rochester, NY: Social Science Research Network. https://papers.ssrn.com/abstract=2584245.

Kolbert, Elizabeth. 2007. *Field Notes from a Catastrophe: Man, Nature, and Climate Change*. New York: Bloomsbury Publishing USA.

———. 2014. *The Sixth Extinction: An Unnatural History*. New York: Henry Holt and Company.

Koniaris, Marios, Ioannis Anagnostopoulos, and Yannis Vassiliou. 2014. "Legislation as a Complex Network: Modelling and Analysis of European Union Legal Sources." In *JURIX*, 143–52.

———. 2018. "Network Analysis in the Legal Domain: A Complex Model for European Union Legal Sources." *Journal of Complex Networks* 6 (2): 243–68. https://doi.org/10.1093/comnet/cnx029.

Kopp, Robert E., and Bryan K. Mignone. 2012. "The U.S. Government's Social Cost of Carbon Estimates after Their First Two Years: Pathways for Improvement." *Economics* 6: 2012–15. https://doi.org/10.5018/economics -ejournal.ja.2012-15.

Kotchen, Matthew J., and Erin T. Mansur. 2016. "Correspondence: Reassessing the Contribution of Natural Gas to US CO_2 Emission Reductions since 2007." *Nature Communications* 7 (March): 10648. https://doi.org/10.1038 /ncomms10648.

KPMG. 2018. "Autonomous Vehicles Readiness Index—Assessing Countries' Openness and Preparedness for Autonomous Vehicles." https://eu-smartcities .eu/sites/default/files/2018-02/avri.pdf.

Krebs, Teri S., and Pål-Ørjan Johansen. 2013. "Over 30 Million Psychedelic Users in the United States." *F1000Research* 2 (March). https://doi.org/10 .12688/f1000research.2-98.v1.

Kreuz, Sebastian, and Felix Müsgens. 2017. "The German Energiewende and Its Roll-Out of Renewable Energies: An Economic Perspective." *Frontiers in Energy* 11 (2): 126–34. https://doi.org/10.1007/s11708-017-0467-5.

Krugman, Paul. 1996. "A Country Is Not a Company." *Harvard Business Review*, January. https://hbr.org/1996/01/a-country-is-not-a-company.

———. 2018. "The Depravity of Climate-Change Denial." *New York Times*, November 26.

Kupers, Roland, ed. 2014. *Turbulence - A Corporate Perspective on Collaborating for Resilience*. Amsterdam: Amsterdam University Press. https://doi .org/10.26530/OAPEN_477310.

Kupers, Roland, Albert Faber, and Annemarth Idenburg. 2015. "Wie is de Wolf? Een systeemblik op de Nederlandse Energietransitie." http://rolandkupers .iutest.nl/wp-content/uploads/2015/11/Wie-is-de-Wolf_final.pdf.

Lansing, J. Stephen, Stefan Thurner, Ning Ning Chung, Aurélie Coudurier-Curveur, Çağil Karakaş, Kurt A. Fesenmyer, and Lock Yue Chew. 2017. "Adaptive Self-Organization of Bali's Ancient Rice Terraces." *Proceedings of the National Academy of Sciences* 114 (25): 6504–9. https://doi.org/10.1073 /pnas.1605369114.

Levitin, Daniel. 2017. *A Field Guide to Lies and Statistics: A Neuroscientist on How to Make Sense of a Complex World*. London: Viking.

Lindblom, Charles E. 1959. "The Science of 'Muddling Through.'" *Public Administration Review* 19 (2): 79–88.

———. 1968. *The Policy-Making Process*. Upper Saddle River, NJ: Prentice-Hall.

Lynas, Mark. 2008. *Six Degrees: Our Future on a Hotter Planet*. National Geographic.

Machiavelli, Niccolò. 1532. *The Prince*. Translated by W. K. Marriott (1908). http://www.constitution.org/mac/prince06.htm.

Mann, Charles C. 2018. *The Wizard and the Prophet: Two Remarkable Scientists and Their Dueling Visions to Shape Tomorrow's World*. New York: Knopf Doubleday Publishing Group.

Mann, Michael E. 2012. *The Hockey Stick and the Climate Wars: Dispatches from the Front Lines*. New York: Columbia University Press.

Marshall, Kristin N., N. Thompson Hobbs, and David J. Cooper. 2013. "Stream Hydrology Limits Recovery of Riparian Ecosystems after Wolf Reintroduction." *Proceedings of the Royal Society of London B: Biological Sciences* 280 (1756): 20122977. https://doi.org/10.1098/rspb.2012.2977.

Martínez, Alberto A. 2005. "Handling Evidence in History: The Case of Einstein's Wife." *School Science Review* 86 (316): 49–56.

Mazzucato, Mariana. 2014. "The Innovative State." *Foreign Affairs*, December 15. https://www.foreignaffairs.com/articles/americas/2014-12-15/innovative-state.

———. 2018. *The Entrepreneurial State: Debunking Public vs. Private Sector Myths*. Penguin. https://www.penguin.co.uk/books/305469/the-entrepreneurial-state/.

McKenzie, Evan, and Jay Ruby. 2002. "Reconsidering the Oak Park Strategy: The Conundrums of Integration." *Midwest Political Science Association*. http://astro.temple.edu/~ruby/opp/3qrpt02/finalversion.pdf.

McKibben, Bill. 1989. *The End of Nature*. New York: Random House Trade Paperbacks.

———. 2013. *Oil and Honey: The Education of an Unlikely Activist*. London: Macmillan.

Meng, Kyle C. 2016. "Estimating Path Dependence in Energy Transitions." Working Paper 22536. National Bureau of Economic Research. https://doi.org/10.3386/w22536.

Merton, Robert K. 1936. "The Unanticipated Consequences of Purposive Social Action." *American Sociological Review* 1 (6): 894–904. https://doi.org/10.2307/2084615.

Middleton, Arthur. 2014. "Is the Wolf a Real American Hero?" *New York Times*, March 9, Opinion. https://www.nytimes.com/2014/03/10/opinion/is-the-wolf-a-real-american-hero.html.

Mishina, Kazuhiro. 1999. "Learning by New Experiences: Revisiting the Flying Fortress Learning Curve." In *Learning by Doing in Markets, Firms, and Countries*, 145–84. Chicago: University of Chicago Press. http://www.nber.org/chapters/c10232.pdf.

Mohlin, Kristina, Jonathan R. Camuzeaux, Adrian Muller, Marius Schneider, and Gernot Wagner. 2018. "Factoring in the Forgotten Role of Renewables in CO_2 Emission Trends Using Decomposition Analysis." *Energy Policy* 116 (May): 290–96. https://doi.org/10.1016/j.enpol.2018.02.006.

Moor, Robert. 2016. "What Happens to American Myth When You Take the Driver Out of It?" *New York Magazine*, October 16.

Morton, Oliver. 2015. *The Planet Remade: How Geoengineering Could Change the World*. Princeton, NJ: Princeton University Press.

Müller, Sven, and Johannes Rode. 2013. "The Adoption of Photovoltaic Systems in Wiesbaden, Germany: Economics of Innovation and New Technology" *Economics of Innovation and New Technology* 22 (5): 1–17. https://doi.org/10.1080/10438599.2013.804333.

National Highway Traffic Safety Administration. 2015. "Critical Reasons for Crashes Investigated in the National Motor Vehicle Crash Causation Survey." *Traffic Safety Facts*, February. https://crashstats.nhtsa.dot.gov/Api/Public/ViewPublication/812115.

Nelson, Richard R., and Sidney G. Winter. 2002. "Evolutionary Theorizing in Economics." *Journal of Economic Perspectives* 16 (2): 23–46. https://doi.org/10.1257/0895330027247.

Newell, Richard G, Adam B. Jaffe, and Robert N Stavins. 1998. "The Induced Innovation Hypothesis and Energy-Saving Technological Change." National Bureau of Economic Research, Cambridge MA.

New York Times. 2018. "Wake Up, World Leaders. The Alarm Is Deafening." *New York Times*, October 12, Opinion.

———. 2019. "You Call That Meat? Not So Fast, Cattle Ranchers Say." *New York Times*, February 9, 2019. https://www.nytimes.com/2019/02/09/technology/meat-veggie-burgers-lab-produced.html.

Nordhaus, William D. 2007. "A Review of the *Stern Review on the Economics of Climate Change*." *Journal of Economic Literature* 45 (3): 686–702. https://doi.org/10.1257/jel.45.3.686.

Nour, Matthew M., Lisa Evans, and Robin L. Carhart-Harris. 2017. "Psychedelics, Personality and Political Perspectives." *Journal of Psychoactive Drugs* 49 (3): 182–91. https://doi.org/10.1080/02791072.2017.1312643.

NTSB. 2018. *Preliminary Report Highway: HWY18MH010*. Accident report HWY18MH010. National Transportation Safety Board. https://www.ntsb.gov/investigations/AccidentReports/Pages/HWY18MH010-prelim.aspx.

Nutt, David J., Leslie A. King, and Lawrence D. Phillips. 2010. "Drug Harms in the UK: A Multicriteria Decision Analysis." *The Lancet* 376 (9752): 1558–65. https://doi.org/10.1016/S0140-6736(10)61462-6.

OECD NEA. 2016. "Costs of Decommissioning Nuclear Power Plants." NEA No. 7201. OECD NEA. http://www.oecd-nea.org/ndd/pubs/2016/7201-costs-decom-npp.pdf.

Olson, J. C. (Jerry Corrie). (1976). *Price as an informational cue: effects on product evaluations*. University Park, Pa.: College of Business Administration, Pennsylvania State University.

Ostrom, Elinor. 1990. *Governing the Commons: The Evolution of Institutions for Collective Action*. Cambridge: Cambridge University Press.

———. 2009. "A Polycentric Approach for Coping with Climate Change." Policy Research Working Paper 5095. World Bank. http://documents.worldbank.org/curated/en/480171468315567893/pdf/WPS5095.pdf.

Page, Scott E. 2006. "Path Dependence." *Quarterly Journal of Political Science* 1 (1): 87–115.

Pangallo, Marco, Torsten Heinrich, and J. Doyne Farmer. 2019. "Best Reply Structure and Equilibrium Convergence in Generic Games." *Science Advances* 5 (2): eaat1328. https://doi.org/10.1126/sciadv.aat1328.

Perez, Carlota. 2010. "Technological Revolutions and Techno-Economic Paradigms." *Cambridge Journal of Economics* 34 (1): 185–202. https://doi.org/10.1093/cje/bep051.

Pimm, S. L., C. N. Jenkins, R. Abell, T. M. Brooks, J. L. Gittleman, L. N. Joppa, P. H. Raven, C. M. Roberts, and J. O. Sexton. 2014. "The Biodiversity of Species and Their Rates of Extinction, Distribution, and Protection." *Science* 344 (6187): 1246752. https://doi.org/10.1126/science.1246752.

Pinker, Steven. 2012. *The Better Angels of Our Nature: Why Violence Has Declined.* Penguin Books. https://www.amazon.com/Better-Angels-Our-Nature-Violence/dp/0143122010.

Pollan, Michael. 2018. *How to Change Your Mind: What the New Science of Psychedelics Teaches Us about Consciousness, Dying, Addiction, Depression, and Transcendence.* New York: Penguin Press.

Porter, Eduardo. 2015. "Behind Drop in Oil Prices, Washington's Hand." *New York Times*, January 20, Economy. https://www.nytimes.com/2015/01/21/business/economy/washingtons-role-in-oil-prices-recent-fall.html.

PQ Media. 2018. "PQ Media's Global Advertising and Marketing Revenue Forecast 2018–22." PQ Media. https://www.pqmedia.com/product/global-advertising-marketing-revenue-forecast-2018-22/.

Pratson, Lincoln F., Drew Haerer, and Dalia Patiño-Echeverri. 2013. "Fuel Prices, Emission Standards, and Generation Costs for Coal vs Natural Gas Power Plants." *Environmental Science and Technology* 47 (9): 4926–33. https://doi.org/10.1021/es4001642.

Prigogine, Ilya, and Isabelle Stengers. 1984. *Order Out of Chaos: Man's New Dialogue with Nature.* New York: Bantam Books.

Radden Keefe, Patrick. 2017. "The Family That Built an Empire of Pain." *New Yorker*, October 30. https://www.newyorker.com/magazine/2017/10/30/the-family-that-built-an-empire-of-pain.

Radnofsky, Caroline. 2016. "Is Gun Violence in the US Infectious?" *Al Jazeera*, July 31. https://www.aljazeera.com/indepth/features/2016/07/gun-violence-infectious-160731083342582.html.

Ramalingam, Ben. 2013. *Aid on the Edge of Chaos: Rethinking International Cooperation in a Complex World.* Oxford: Oxford University Press.

Rand, David G., Joshua D. Greene, and Martin A. Nowak. 2012. "Spontaneous Giving and Calculated Greed." *Nature* 489 (September): 427.

Reinhart, Carmen M., and Kenneth S. Rogoff. 2009. *This Time Is Different: Eight Centuries of Financial Folly.* Princeton, NJ: Princeton University Press.

Ricke, Katharine, Laurent Drouet, Ken Caldeira, and Massimo Tavoni. 2018. "Country-Level Social Cost of Carbon." *Nature Climate Change* 8 (10): 895–900. https://doi.org/10.1038/s41558-018-0282-y.

Riet, Maarten van 't, and Arjen Lejour. 2014. "Optimal Tax Routing: Network Analysis of FDI Diversion." Discussion Paper. CPB (Netherlands Bureau for Policy Analysis). https://papers.ssrn.com/sol3/papers.cfm?abstract_id =2958027#.

Rigg, Jonathan, and Lisa Reyes Mason. 2018. "Five Dimensions of Climate Science Reductionism." *Nature Climate Change* 8 (12): 1030. https://doi.org /10.1038/s41558-018-0352-1.

Rockström, Johan, Will Steffen, Kevin Noone, Åsa Persson, F. Stuart Chapin, Eric F. Lambin, Timothy M. Lenton, et al. 2009. "A Safe Operating Space for Humanity." *Nature* 461 (7263): 472–75. https://doi.org/10.1038/461472a.

Rong, Yaoguang, Yue Hu, Anyi Mei, Hairen Tan, Makhsud I. Saidaminov, Sang Il Seok, Michael D. McGehee, Edward H. Sargent, and Hongwei Han. 2018. "Challenges for Commercializing Perovskite Solar Cells." *Science* 361 (6408): eaat8235. https://doi.org/10.1126/science.aat8235.

Rosenthal, Elisabeth. 2008. "By 'Bagging It,' Ireland Rids Itself of a Plastic Nuisance." *New York Times*, January 31.

Rotman, David. 2019. "The Economic Argument behind the Green New Deal." *MIT Technology Review.* 2019. https://www.technologyreview.com/s /613341/the-economic-argument-behind-the-green-new-deal/.

Sabel, Charles F., and Jonathan Zeitlin. 2012. *Experimentalist Governance.* Oxford University Press. https://doi.org/10.1093/oxfordhb/9780199560530 .013.0012.

Salehi-Isfahani, Djavad. 2017. "From Energy Subsidies to Universal Basic Income: Lessons from Iran." Economic Research Forum (ERF), November 19. https:// theforum.erf.org.eg/2017/11/19/energy-subsidies-universal-basic-income -lessons-iran/.

Schaffer, J, and Sebastian Brun. 2015. "Beyond the Sun—Socioeconomic Drivers of the Adoption of Small-Scale Photovoltaic Installations in Germany." *Energy Research and Social Science* 10 (November): 220–27. https://doi.org /10.1016/j.erss.2015.06.010.

Scheffer, Marten. 2009. *Critical Transitions in Nature and Society.* Princeton, NJ: Princeton University Press.

Schelling, Thomas C. 1969. *Neighborhood Tipping.* Harvard Institute of Economic Research, Harvard University, Cambridge, MA.

———. 1971. "Dynamic Models of Segregation." *Journal of Mathematical Sociology* 1 (2): 143–86.

———. 2006. *Micromotives and Macrobehavior.* New York: W. W. Norton.

Schneider, Stephen H. 2009. *Science as a Contact Sport: Inside the Battle to Save Earth's Climate.* Washington DC: National Geographic Books.

Schumpeter, Joseph A. 1939. *Business Cycles: A Theoretical, Historical and Statistical Analysis of the Capitalist Process.* New York; Toronto; London: McGraw-Hill Book Company.

Sengupta, Somini. 2018. "The World Needs to Quit Coal. Why Is It So Hard?" *New York Times*, November 24.

Shoup, Donald C. 2005. *The High Cost of Free Parking.* Chicago: Planners Press, American Planning Association.

Simon, Herbert A. 1947. "A Comment on 'The Science of Public Administration.'" *Public Administration Review* 7 (3): 200–203. https://doi.org/10.2307/972716.

Sivaram, Varun. 2017. "Unlocking Clean Energy." *Issues in Science and Technology* 33 (2): 31–40.

———. 2018. *Taming the Sun*. Cambridge, MA: MIT Press. https://mitpress.mit.edu/books/taming-sun.

Slutkin, Gary. 2015. "Violence Is a Contagious Disease." Workshop summary. National Academy of Sciences. http://cureviolence.org/wp-content/uploads/2015/05/Violence-is-a-Contagious-Disease.pdf.

Smith, Bryant Walker. 2016. "How Governments Can Promote Automated Driving." SSRN Scholarly Paper ID 2749375. Rochester, NY: Social Science Research Network. https://papers.ssrn.com/abstract=2749375.

Smith, Christopher J., Piers M. Forster, Myles Allen, Jan Fuglestvedt, Richard J. Millar, Joeri Rogelj, and Kirsten Zickfeld. 2019. "Current Fossil Fuel Infrastructure Does Not Yet Commit Us to 1.5 °C Warming." *Nature Communications* 10 (1): 101. https://doi.org/10.1038/s41467-018-07999-w.

Stern, Nicholas. 2016. "Economics: Current Climate Models Are Grossly Misleading." *Nature News* 530 (7591): 407. https://doi.org/10.1038/530407a.

Sterner, Thomas, Edward B. Barbier, Ian Bateman, Inge van den Bijgaart, Anne-Sophie Crépin, Ottmar Edenhofer, Carolyn Fischer, et al. 2019. "Policy Design for the Anthropocene." *Nature Sustainability* 2 (1): 14–21. https://doi.org/10.1038/s41893-018-0194-x.

Stoknes, Per Espen. 2014. "Rethinking Climate Communications and the 'Psychological Climate Paradox.'" *Energy Research and Social Science* 1 (March): 161–70. https://doi.org/10.1016/j.erss.2014.03.007.

Stout, Lynn. 2012. *The Shareholder Value Myth: How Putting Shareholders First Harms Investors, Corporations, and the Public*. San Francisco: Berrett-Koehler.

Thompson, Andrea. 2018. "What's in a Half a Degree? 2 Very Different Future Climates." *Scientific American*, October. https://www.scientificamerican.com/article/whats-in-a-half-a-degree-2-very-different-future-climates/.

Thurner, Stefan, Rudolf Hanel, and Peter Klimek. 2018. *Introduction to the Theory of Complex Systems*. Oxford; New York: Oxford University Press.

Tsafos, Nikos. 2018. "Must the Energy Transition Be Slow? Not Necessarily." CSIS Briefs. Center for Strategic and International Studies. https://csis-prod.s3.amazonaws.com/s3fs-public/publication/181001_Energy_Transistion.pdf.

Tuchman, Barbara. 1986. *Guns of August: The Drama of August 1914*. New York: Random House.

UberATG. 2018. "The Future of Trucking:" *UberATG* (blog), February 1. https://medium.com/@UberATG/the-future-of-trucking-b3d2ea0d2db9.

Umweltbundesamt. 2014. "*Treibhausgasneutrales Deutschland Im Jahr 2050—Studie*." Climate Change 07/2014. Berlin: Umweltbundesamt. https://www.umweltbundesamt.de/publikationen/treibhausgasneutrales-deutschland-im-jahr-2050-0.

UN Environment. 2018. "Inclusive Wealth Report 2018—Executive Summary (IWR)." Nairobi: UN Environment. https://wedocs.unep.org/bitstream /handle/20.500.11822/26776/Inclusive_Wealth_ES.pdf.

Unruh, Gregory C. 2000. "Understanding Carbon Lock-In." *Energy Policy* 28 (12): 817–30. https://doi.org/10.1016/S0301-4215(00)00070-7.

US EPA, OA. 2016. *Retrospective Study of the Costs of EPA Regulations: A Report of Four Case Studies*. Reports and Assessments. US EPA, March 31. https://www.epa.gov/environmental-economics/retrospective-study-costs-epa -regulations-report-four-case-studies.

Van Benthem, Arthur, Kenneth Gillingham, and James Sweeney. 2008. "Learning-by-Doing and the Optimal Solar Policy in California." *Energy Journal* 29 (3): 131–51.

Wagner, Gernot, Tomas Kåberger, Susanna Olai, Michael Oppenheimer, Katherine Rittenhouse, and Thomas Sterner. 2015. "Energy Policy: Push Renewables to Spur Carbon Pricing." *Nature* 525 (7567): 27–29. https://doi.org/10 .1038/525027a.

Wagner, Gernot, and Martin L. Weitzman. 2015. *Climate Shock: The Economic Consequences of a Hotter Planet*. Princeton, NJ: Princeton University Press.

Waldrop, Mitchell M. 1992. *Complexity: The Emerging Science at the Edge of Order and Chaos*. 1st ed. New York: Simon and Schuster.

Walker, Brian. 2009. *The Best Explanation to Resilience*. Stockholm Resilience Centre TV. Stockholm. https://www.youtube.com/watch?v=tXLMeL5nVQk &feature=youtu.be.

Walker, Brian, and David Salt. 2006. *Resilience Thinking: Sustaining Ecosystems and People in a Changing World*. Washington DC: Island Press.

Walras, Léon. 2013. *Elements of Pure Economics*. Abingdon UK: Routledge. https://doi.org/10.4324/9781315888958.

Watts, Duncan J. 2002. "A Simple Model of Global Cascades on Random Networks." *Proceedings of the National Academy of Sciences of the United States of America* 99 (9): 5766–71.

Watts, Duncan, and Steven Strogatz. 1998. "Collective Dynamics of 'Small-World' Networks." *Nature* 393 (April): 440–42.

Weitzman, Martin L. 1998. "Recombinant Growth." *Quarterly Journal of Economics* 113 (2): 331–60.

Wendling, Z. A., Emerson, J. W., Esty, D. C., Levy, M. A., de Sherbinin, A., *et al.* (2018). *2018 Environmental Performance Index*. New Haven, CT: Yale Center for Environmental Law & Policy. https://epi.yale.edu/.

WHO. 2015. *Global Status Report on Road Safety 2015*. Rome, Italy: World Health Organization.

———. 2018. *Global Status Report on Road Safety 2018*. Rome, Italy: World Health Organization. https://www.who.int/violence_injury_prevention/road _safety_status/2018/en/.

Wilkinson, Angela, and Roland Kupers. 2014. *The Essence of Scenarios: Learning from the Shell Experience*. Amsterdam: Amsterdam University Press. http://www.press.uchicago.edu/ucp/books/book/distributed/E/bo18045004 .html.

Wilson, David Sloan. 2016. "Failed Economics: Tyranny of Mathematics and Enslaved by the Wrong Theory." *Evonomics: The Next Evolution of Economics* (blog), January 28. http://evonomics.com/failed-economics -tyranny-of-mathematics-enslaved-wrong-theory/.

Wired. 2015. "Opinion: How the US Embassy Tweeted to Clear Beijing's Air." *Wired*, March 6. https://www.wired.com/2015/03/opinion-us-embassy -beijing-tweeted-clear-air/.

Yang, Andrew. 2018. "Self-Driving Vehicles: What Will Happen to Truck Drivers?" *Evonomics*, October. http://evonomics.com/what-will-happen-to -truck-drivers-ask-factory-workers-andrew-yang/?utm_source=newsletter _campaign=organic.

Yellowstone Park. 2011. "Wolf Reintroduction Changes Ecosystem." *Yellowstone Wildlife Guide Including Grizzly, Wolves, and Bison.* June 21. https://www.yellowstonepark.com/things-to-do/wolf-reintroduction-changes -ecosystem.

Yergin, Daniel. 2011. *The Prize: The Epic Quest for Oil, Money and Power.* New York: Simon and Schuster.

York, Richard, and Julius Alexander McGee. 2016. "Understanding the Jevons Paradox." *Environmental Sociology* 2 (1): 77–87. https://doi.org/10.1080 /23251042.2015.1106060.

Young, Emma. 2017. "How Iceland Got Teens to Say No to Drugs." *The Atlantic*, January 19. https://www.theatlantic.com/health/archive/2017/01 /teens-drugs-iceland/513668/.

Zeitlin, Jonathan. 1995. "Flexibility and Mass Production at War: Aircraft Manufacture in Britain, the United States, and Germany, 1939–1945." *Technology and Culture* 36 (1): 46–79.

ACKNOWLEDGMENTS

My gratitude goes out to Gernot Wagner, with whom I closely collaborated in the early stages of writing. His ideas have influenced—and indeed improved—several arguments in this book.

A warm thank you to the editors Thomas LeBien and Andrew Kinney and to all the excellent staff from Harvard University Press, as well as copy editor Barbara Goodhouse and production editor John Hoey, who have patiently provided support, encouragement, and also robust challenge when it was needed.

Many thanks to Bernise Ang, Maike Böggemann, Arjen Bongard, Paul van der Cingel, Ann Florini, Karen Florini, Mark Foden, Annemarth Idenburg, Maggie Neil, Georgie Passalaris, Shayna Rector Bleeker, Rachel Sinha, Simon Zadek, and Joe Zammit-Lucia for providing feedback on various parts of the book.

The Institute for Advanced Studies of the University of Amsterdam and the Complexity Institute of Nanyang Technological University have provided valuable support and assistance.

Finally, I owe a great debt to my father, who passed away during the writing of this book, for the intellectual curiosity that he inspired over a lifetime.

INDEX

Note: Page numbers followed by *f* or *t* indicate figures or tables, respectively.
Page references including n indicate endnotes.